MICRO-ORGANISMS IN FOODS

2

Sampling for microbiological analysis: Principles and specific applications

The International Commission on

Microbiological Specifications for Foods (ICMSF)

of the International Association of

Microbiological Societies

UNIVERSITY OF TORONTO PRESS

Toronto Buffalo London
Reprinted, with corrections, 1978
Printed in Canada
ISBN 0-8020-2143-3
LC 73-2628

As the production of foodstuffs grows more centralized in larger and larger enterprises, and as foods move internationally on a rapidly increasing scale, there is an urgent need for statistically based sampling plans in order to ensure the quality of what is eaten. Microbiological methods for appraising food quality, no matter how accurate and reproducible, are inadequate without a satisfactory sampling plan. Currently used sampling plans are often highly unsatisfactory, and the level of safety they provide is, for the most part, poorly understood. This book dispels much of the controversy and indecision surrounding microbiological methods of appraisal, and is the first comprehensive publication dealing with statistically based sampling plans, broadly applicable to the field of food microbiology.

The volume is divided into two sections. Part I defines statistical concepts and terms, describes the sampling plans available, outlines the procedures for selecting the best plan, and explains the principles of randomization. Part II describes the practical applications of these principles for various foods, including fish, shellfish, raw meats, processed meats, vegetables, milk and milk products, dried foods, frozen foods, and shelf-stable canned foods.

Fifty-three experienced scientists from twenty-four countries have contributed five years of study and discussion to this important book. *The International Commission on Microbiological Specifications for Foods* is a standing commission of the International Association of Microbiological Societies. It was formed in 1962 in response to the need for internationally acceptable and authoritative decisions on microbiological limits for foods commensurate with public health safety, and particularly for foods in international commerce. Its members are all internationally known food microbiologists, drawn from government laboratories in public health, agriculture, and food technology, and from universities and the food industry.

SPONSORED BY

the International Commission on
Microbiological Specifications for Foods,
of the International Association
of Microbiological Societies

EDITORIAL COMMITTEE

M. Ingram (Chairman), D.F. Bray, D.S. Clark,
C.E. Dolman, R.P. Elliott, and F.S. Thatcher

CONTRIBUTORS

ICMSF *Members*

F.S. Thatcher (Chairman),
D.S. Clark (Secretary-Treasurer), H.E. Bauman,
J.H.B. Christian, C. Cominazzini, C.E. Dolman,
R.P. Elliott, O. Emberger, H.E. Goresline,
Betty C. Hobbs, H. Iida, M. Ingram, K.H. Lewis,
H. Lundbeck, G. Mocquot, G.K. Morris, D.A.A. Mossel,
N.P. Nefedjeva, J.C. Olson, Jr., F. Quevedo,
B. Simonsen, H.J. Sinell

Subcommission Members

Ph.V. Bartl, Z. Bulajić, N.S. de Caruso,
J. Gomez Ruiz, C. Ienistea,
M. Kalember-Radosavljević, S. Mendoza,
M. Šipka, J. Takács, F. Yanc, Z. Zachariev

Consultants

G.G. Anderson, H. Beerens, D.F. Bray, I.W. Burr,
E.F. Drion, I.E. Erdman, E.J. Gangarosa,
A. Hurst, L. Jovanović, J. Liston,
M.S. Lowenstein, D.A. Lyon, Z. Matyas,
C.F. Niven, L. Ormay, H. Pivnick, L. Reinius,
J.M. Shewan, J.H. Silliker, P. Vassiliadis

Contents

Preface

Microbiological methods, no matter how accurate and reproducible, are inadequate to appraise the microbiological quality of food without a satisfactory sampling plan. Sampling plans in current use are often badly understood and highly unsatisfactory, and the level of safety they provide is generally not well known. At present, no comprehensive publication exists dealing with statistically based sampling plans, broadly applicable to the field of food microbiology. The facts that the production of foodstuffs is gradually being centralized in enlarging enterprises, and that foodstuffs are moving in international commerce on a rapidly increasing scale, make the need for such plans still more obvious. The Commission decided to try to fill this gap and, to that end, called upon expert statisticians and commodity specialists to supplement its own areas of competence (see Appendix 2).

The primary aim of this book is instructional: to draw attention to the unsatisfactory nature of casual sampling, and to provide information necessary to improve the situation. The immediate objectives are: to describe what different sampling plans can achieve in principle; to establish their significance for the interpretation of the results of microbiological testing of foods; and to offer guidance as to the choice of appropriate sampling plans in different situations.

The cost of sampling and testing and inadequate awareness of the principles involved nearly always preclude the use of an ideal sampling plan, and lead to the use of sampling procedures that are far from satisfactory in both nature and scope. Such a situation seems likely to produce an unjustified feeling of security in those responsible for the interpretation of the results. Improved sampling plans as indicated in this book will give a basis for statistically valid and realistic judgment of the microbiological quality of foods under test.

The Commission has limited its scope both in the previously published book, and in the present text, to the range of competence of its members and advisers. Thus neither includes the fields of parasitology, virology, and mycology. Incomplete survey data imposed additional limitations. The criteria proposed herein were based primarily on the knowledge of experts in the various commodity fields. They will apply to many situations, but must be tempered with good judgment in particular instances. The Commission intends to collect further information on these matters, which will probably make imperative a revision of this text in the relatively near future.

The scope and limitations, as outlined above, indicate that the book is intended primarily for people dealing with the microbiological safety of food moving in international and domestic commerce. National food control and public health authorities, food manufacturers, international organizations, and teachers in food microbiology should also find it useful.

The substance of the book falls into two parts: the principles of sampling (Part I), and the application of these principles to suggest sampling plans for various foods (Part II). Part I defines the statistical concepts and terms used in the book, describes the kind of sampling plans available and their purposes, outlines the procedures for selecting the best plan for the food and organism in question, and explains the principles of randomization. Part II describes practical applications of the principles set out in Part I and treats the procedures for collecting and handling sampling units for various food categories.

The text ends with general conclusions and a glossary of special terms.

The text is the result of five years of study by ICMSF members and consultants, involving workshops held in Dubrovnik, Yugoslavia (1969), Mexico City, Mexico (1970), Opatija, Yugoslavia (1971), Langford, England (1972), and Ottawa, Canada (1973). The project was proposed and planned by the chairman, F.S. Thatcher, after initial phases of collaboration with the chairman of the Statistics Committee, D.F. Bray. Major segments of the work were first studied by subcommittees (see Appendix 3) and then discussed in plenary sessions. Results were compiled as minutes by the secretary, D.S. Clark, and later drafted into book form by the chairman, F.S. Thatcher. Subsequent compilation and rewriting was done by the various subcommittees. Editing was the responsibility of an editorial committee under Chairman M. Ingram.

The group represented a range of opinions and principles, whose expression required constructive compromise. For example, the impracticability at present of insisting that *Salmonella* be undetectable in some raw meats had to be reconciled with the reluctance to sanction the presence of a pathogen in any food. The sampling plans described herein are not inflexible standards. If applied with increasing stringency they should permit improvement in microbiological quality without making necessary the condemnation, in prohibitive proportion, of important foodstuffs.

NOTICE

The main aims of this book are to show what sampling plans can do and what are their advantages over casual sampling procedures. The examples set out in Part II are only intended to illustrate the organisms important in the different foods and how, for each, limits might be set. The Commission has not decided whether there is a sufficient need to apply all these limits. Moreover, the actual limits chosen are based on information readily available to the Commission, which is mainly derived from a restricted range of countries and conditions; hence the suggested values are likely to need revision if applied to different circumstances. The values indicated are certainly not intended as hard and fast limits for universal application.

ACKNOWLEDGMENTS

Sincere thanks are extended to the many scientists who contributed to this book, in particular: Dr D.F. Bray, Professor I.W. Burr, and Dr E.F. Drion, who guided the Commission on statistical questions; Mr G.G. Anderson, Dr J. Liston, Dr C.F. Niven, and Dr J.M. Shewan, whose advice and assistance were much appreciated; the chairmen of the various subcommittees; and also to the staff of the National Science Library of Canada, for confirming the accuracy of the reference citations.

The Commission is most grateful for generous financial sponsorship without which the work would have been impossible: from the United States Department of Health, Education and Welfare, Public Health Service, Health Services and Mental Health Administration, Center for Disease Control; the World Health Organization; Canada Department of National Health and Welfare, Health Protection Branch; and various companies within the food industry (see Appendix 4). This assistance does not, of course, constitute endorsement of the findings and views expressed herein. Finally, thanks are also expressed to the respective national governments, universities, and private companies for supporting the participation of their staff in the work of the Commission, of which the present text is but one result.

ABBREVIATIONS OF SOCIETIES AND AGENCIES

AFDOUS	Association of Food and Drug Officials of the United States
AOAC	Association of Official Analytical Chemists
APHA	American Public Health Association
FAO	Food and Agriculture Organization of the United Nations
IAEA	International Atomic Energy Agency
IAMS	International Association of Microbiological Societies
ICMSF	International Commission on Microbiological Specifications for Foods
IDF	International Dairy Federation
ISO	International Standards Organization
NAS	National Academy of Sciences of the United States
NRC	National Research Council of the United States
PHS	Public Health Service of the United States
UNICEF	United Nations International Children's Emergency Fund
USFDA	United States Food and Drug Administration
USDHEW	United States Department of Health Education and Welfare
WHO	World Health Organization of the United Nations

PART I PRINCIPLES

1

Concepts of probability and sampling

Consider a trial or a test, the outcome of which is doubtful, such as a test for the presence of an organism in a food. After the standard procedure has been followed, the test either shows the presence of the organism or it does not; thus we have a positive or a negative result. We have resolved the pre-test uncertainty.

If there were relatively many of the organisms present in the food sampled, we would expect relatively many such tests to yield a positive result. But if there were relatively few organisms present, we would expect that relatively few such tests would yield a positive result. In these two cases, the 'probability' of a positive result would be respectively high and low.

A PROBABILITY

The probability of a positive result is, in fact, the 'long-run' proportion of times a positive result occurs out of all the times we test the food. Thus, if a positive result occurs 112 times in 1000 tests, we estimate the probability to be $112/1000 = 0.112$, while if it occurs 914 times in 1000 trials, we estimate the probability to be $914/1000 = 0.914$.

The word 'estimate' is used because, if we were to run 1000 trials again on the same material and using the same procedure, we could not be sure of observing 112 and 914 positive results, respectively. But the results should be close to these, because 1000 is a large number of trials. Thus the estimated probability for a positive result (or any other outcome in some other type of test) is the proportion of times the outcome did occur among the trials or tests actually made. A probability can be anywhere from 0 to 1. It will be zero if the organism is absent from the food (assuming a test procedure that will detect any organism present), and 1 if every one of a large number of tests provides a positive result.

What does an observed proportion of trials such as 0.112 estimate? Suppose we divided the entire lot of food into small sample units, say perhaps 10,000,000 one-gram units, and then went through the test procedure on every such sample unit. Suppose 1,051,200 gave positive results. Then the ratio 1,051,200/10,000,000 (the actual proportion of positives) = 0.10512 is the measure of the probability. But what kind of measure? This is no longer an 'estimate' of the probability of a positive. Instead, it is the 'true probability' or 'population probability.' Of course, this approach is not practical, because of the test time needed, and because there would be no food left to eat! But it is useful to have this concept in mind. The 'population probability' determines the kind of sample probabilities or estimated probabilities we may expect from a given number of sample units examined. If that number is low, the estimated probability is not likely to be precise. In general, we never know the true or population probability, but have only estimates of it. But we do know that the larger the number of units included in the sample of the population, the closer the estimated probability is likely to be to the population probability.

B POPULATION AND SAMPLE OF THE POPULATION

The preceding section introduced the two basic concepts: (i) population as a whole and (ii) sample of the population. In relation to results of an experimental test, these are respectively: the whole set of results of the test made on each and every unit in the lot of food, and the partial collection of such results, derived only from the group of sample units actually examined. In terms of standard plate counts, for example, these concepts would be represented by (i) all the counts that could be made by examining every unit in the lot; and (ii) those counts actually made on the few sample units examined. Statisticians use the word 'sample' for the group of units which is withdrawn to estimate the character of a population, while an analyst or bacteriologist would call any one of these units a 'sample.' To try to minimize this confusion, we distinguish between 'the sample of the population' and the 'sample units' of which it is composed.

C DISTRIBUTIONS OF COUNTS

Any series of measurements or counts can be analysed by tallying, that is, by taking each number in turn, and tallying it in whatever numerical interval it falls. For example, we might have classes for plate counts of 8000–9999, 10,000–11,999, 12,000–13,999, etc. Then we tally each count in the class to which it belongs. This arrangement of observed data into classes is called a 'frequency distribution.' In order to find the estimated probability of a count between, say, 10,000 and 14,000, we add the frequencies for those classes lying between these limits, namely 10,000–11,999 and 12,000–13,999, and then divide the sum of these two frequencies by the total frequency for all the classes. For example, this might be 15 divided by 300, which would be 0.050. In general, the larger the total number of such counts tallied, the more the frequencies of the classes tend to smooth out on plotting and appear like a curve.

D CHOOSING THE SAMPLE UNITS

Chapter 2 describes how to choose the material to be tested from the total amount in the lot or shipment. The important point is to avoid bias, so that the sample of the population may represent the lot as well as possible. Random choice is one way of achieiving this. Thus, if we can think of the lot as made up of a population of 10 gram blocks called 'sample units,' and we decide upon a sample from this population of 10 such units, then we should choose these units in such a way that each sample unit in the lot has the same chance of being included among the sample units chosen. At the very least, we should try to draw test material from all the various parts of the lot.

E THE SAMPLING PLAN

The sample units so drawn will, after examination, yield results which will be compared with certain criteria to reach a decision as to whether the entire lot should be accepted or rejected (see Section H, page 8 for explanation of rejection). The particular choice of sampling procedure and the decision criteria is called the 'sampling plan.' Proper drawing of samples is essential, if the decision-making criteria of the sampling plan is to give unbiased results. By sampling randomly we are able to reduce the risks of wrong decisions.

A simple example of a sampling plan follows: Take ten sample units of a food from the lot and subject them to appropriate laboratory procedure to test for the presence or absence of a microorganism. If two or fewer of the ten sample units show presence of the organism, that is, give a 'positive' result, then the whole lot of food is acceptable (relative to this organism). But if three or more give a positive result, the whole lot is to be rejected. This plan is described by $n = 10$ (number of sample units drawn) and $c = 2$ (maximum allowable number of positive results).

F THE OPERATING CHARACTERISTIC CURVE

Section E described what we mean by a sampling plan, and gave an example of a simple sampling plan based upon positive and negative indications of a microorganism. This plan was described by the two numbers $n = 10, c = 2$. If we are going to use this plan, we want to know what it will do for us. How discriminating is it? It is possible when using such a plan that it will sometimes, though rarely, accept a relatively poor lot. On the other hand, it is also possible, though rarely, for it to reject a relatively good lot. There is no way to avoid some degree of risk in each acceptance and in each rejection, unless we test the entire lot, in which case no edible food will be left. We can make these risks smaller by testing more sample units, i.e. by using a larger sample size (larger n). In fact, we can reduce the risks to any desired level by making n sufficiently large. But in practice we usually have to seek a compromise between (a) large n (many sample units) and small risks, and (b) small n (few sample units) and large risks.

Here we often use an 'operating characteristic' curve (abbreviated as OC curve). Such a curve has two scales. The vertical scale gives the probability of acceptance, P_a; that is, P_a is the expected proportion of times that the results will indicate 'pass the lot,' out of the number of times a lot of this given quality is sampled for a decision. The horizontal scale shows a measure of a lot quality. The commonest such measure of lot quality is the true probability or percentage of times that a test on a sample unit from the lot of this quality would show an unsatisfactory result (e.g., a 'positive,' or a count above some number m, see Section B, page 20). This is called p, and might be anywhere between 0 and 100%. This probability is commonly expressed as percentage of 'defectives.' Let us see how the probability runs for $n = 10$, $c = 2$ (the calculations are carried out by using the so-called binomial distribution):

$p\%$	0	10	20	30	40	50	60
P_a	1.00	0.93	0.68	0.38	0.17	0.05	0.01

Figure 1 shows the full curve (cf. Table 3). We see that the greater the probability (p) that a test unit from a given lot yields a positive result, the lower is the probability of acceptance (P_a) of that lot. If, for example, we set a limit of 20% defectives (i.e. $p = 20\%$), then $P_a = 0.68$. This means that on 68 of every 100 occasions when we sample a 20% defective lot, we may expect to have 2 or fewer of the 10 tests showing the presence of the organism and thus calling for 'acceptance,' while on 32

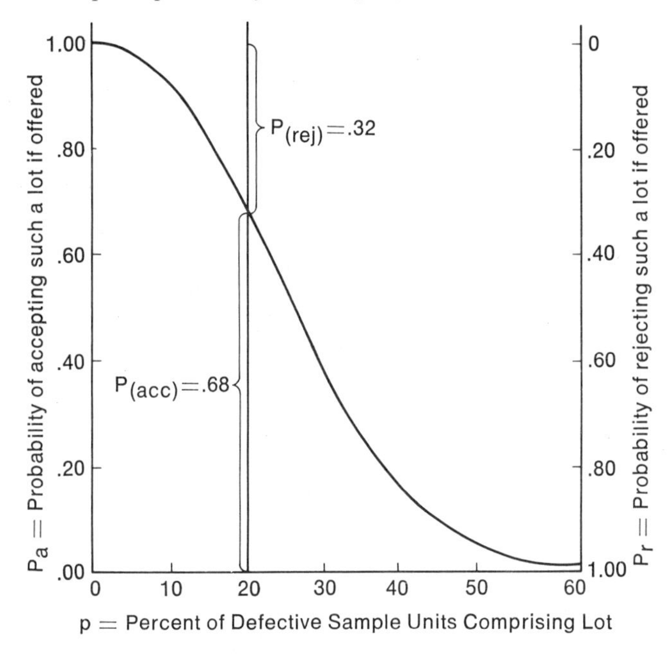

Figure 1 The operating characteristic curve for $n = 10$, $c = 2$, i.e., the probability of accepting lots, in relation to the proportion defective among the sample units comprising the lots

of every 100 there will be three or more positives, calling for non-acceptance. But if $p = 10\%$ (i.e., 10% positive units in the lot), such lots will be accepted 0.93 of the time; or if $p = 40\%$, such lots will be accepted 0.17 of the time. Thus lots with 10% of the sample units defective will be accepted most of the time, but if 40% defective, rather seldom.

G CONSUMER RISK AND PRODUCER RISK

Since decisions to accept or reject lots are made on samples drawn from these lots, occasions will arise when the sample results do not reflect the true condition of the lot. The 'producer's risk' describes the probability that an 'acceptable' lot if offered will be falsely rejected. The 'consumer's risk' describes the probability that a 'bad' lot when offered will be falsely accepted. 'Consumer's risk,' for the purpose of this text, is considered to be the probability of accepting a lot whose actual microbial content is substandard as specified in the plan, even though the determined values indicated acceptable quality. This is expressed by the probability of acceptance (P_a) as given in Tables 2–5. The 'producer's risk' (the obverse of 'consumer's risk') is expressed by $1 - P_a$.

H ACCEPTANCE AND REJECTION

We have been speaking of acceptance or rejection of a lot, on the basis of a sampling plan associated with some particular microbiological test. This judgment, of course, applies only to the purpose for which that test (or several such tests) was performed. A food unsuitable for one purpose may still be suitable for another; for example, if 'rejected' for humans it might still be suitable for animals. Or a rejected food might even, if sorted to remove objectionable material, or if re-processed, be so improved as to pass the test and become acceptable for the original purpose. Normally, therefore, a rejected lot will simply be withheld while the responsible authority decides what to do with it: to return it to the producer, order re-processing, forbid its use for human consumption, or order its destruction, according to circumstances. Throughout this text, the terms 'accept' and 'reject' are used in this limited sense.

2

Principles of drawing samples

Sampling is necessary because it is usually not feasible to subject an entire lot to microbiological examination. Examination of every subunit in the lot would be destructive and too expensive. Results from the portion of the lot represented by the sample are used to draw conclusions about the whole (the lot). This involves the concepts of 'population,' 'sample' (consisting of the 'sample units'), and the associated 'probability,' discussed in Chapter 1.

A WHAT IS A 'LOT'?

Ideally a lot is a quantity of food produced and handled under uniform conditions. In practice, this usually means food produced within a limited period of time. The less uniform the conditions of production, the shorter that period of time should be. To apply such a principle implies knowledge about the uniformity of conditions.

On the other hand, when the supplier is remote from the place where the examination is being made, the examiner may not know how uniform production and handling has been or what should constitute a lot. In this situation, it is helpful if suppliers can be persuaded to give identifiable lot numbers to batches of food produced over short times (e.g., a day, or half a day). Under good manufacturing practice, a lot number is a code which allows identification of the food in relation to some aspect of production, such as the end product of a particular mixing operation, of different formulations, or of a time interval in the production schedule (for example, hourly, daily, a work shift, an interval between 'breaks' for the working crew, an interval between 'clean-up' operations, or other time interval proposed by the Quality Control or Production

Management). Such time intervals are often associated with a particular production line, a retort, or some other critical item of equipment.

Such a 'lot' may include large numbers of small packages, boxes, bags, drums or cartons, all marked with the same lot number; or, if a bulk product, one or more truck- or car-loads. A consignment may contain a number of lots. Where feasible, each lot should be sampled and analysed independently in the interests both of ensuring safety and of minimizing the extent of any rejection if analysis shows any lot to be unacceptable.

When a consignment consists of a very large amount of food not identifiable as separate lots, then a judgment needs to be made on the quantity to be tested as though it constituted a lot (for example, each car-load, or a large container, or many small packages). Small consignments, all containing the same kind of food, and not identifiable by lot number, can be treated only as individual lots, i.e., such a consignment is equivalent to a lot for analytical purposes.

Examination of samples from a series of identifiable lots at regular intervals will soon provide an objective indication of the uniformity of the product, and hence of the conditions of production and handling and the required frequency of acceptance sampling. This is, of course, the basis for the suggestion below (Section K, page 57) that less frequent examinations may be acceptable when there is a continuing record of good performance by the supplier.

B WHAT IS A 'REPRESENTATIVE' SAMPLE?

A 'representative' sample is one whose condition is as similar as possible to that of the lot from which it is drawn. Ideally, one seeks to draw a collection of sample units such that the quality of the sample they constitute is neither better nor worse than that of the lot as a whole.

How then should one proceed to draw a representative sample? The main objective is to avoid bias and to draw a sufficient number of sample units. Sampling at random is the universally recognized way of avoiding bias. It is safer than consciously trying to draw the sample units from the various parts of a lot. Chance does the work. For this, the units (cartons or containers, particular weights of solid, or volumes of liquid) are drawn by the use of random sampling numbers. There is, of course, no guarantee that a sample chosen by random numbers (see Section C, below) has characteristics identical with those of the lot. But one can be confident that the sample was chosen in an approved and unbiased manner.

It is also possible, and desirable, to use a 'stratified' random sampling approach, drawing a random sample of a given number of sample units from each 'stratum' (e.g. each sublot, or carton, or from a chosen number of such sublots or cartons). The objective is to give each small bit of material within each stratum an equal chance to be in the total sample. Such stratification is a device for controlling known sources of variation, and may be used where one has prior knowledge that the consignment is potentially not of uniform quality. This may arise if portions of the consignment are shipped in different vehicles or in different holds of a vessel; or if it is known that the consignment is really composed of several lots, representing, for example, different days of production from the same plant or different plants of the same company. The results for different strata should be assessed separately and then pooled if they appear to be homogeneous.

Heterogeneity within sample units is a phenomenon analogous to stratification. For some food products, different areas within a sample unit are expected to have different quality characteristics. Cans of frozen liquids, for example, are exposed to a temperature gradient during freezing and thus bacteria are concentrated at the centre. Such factors should be carefully explored in sufficient detail to determine the exact method by which to draw the sample (see Chapter 1).

C USE OF TABLE OF RANDOM NUMBERS

The most unbiased and approved method of random sampling is the use of a table of random numbers, which consists of columns of single digits 0, 1, 2, ..., 9 in random order (Table 1). These have been generated by a calculating machine at random, so that, in any column, the digit which comes next is completely independent of the preceding digits in this or other columns. (At each such place the probability for each one of the ten digits to appear next is 0.1.)

To use a table of random numbers or digits, first give consecutive numbers to each package, or container, or other unit, in any convenient order. Then pick a page of the table of random numbers (Table 1), e.g., by numbered cards in a box. Next, without looking, bring a pencil point down on some place on the page. The digit nearest to the point is the first digit. Take as many columns as needed to number all the packages in the lot. Note that the column number is indicated at the top of the table and that each column consists of a vertical row of one-digit numbers. The columns are arranged in groups of 5 to facilitate identification.

To clarify this procedure, take the following example. Suppose a lot of 600 packages has been numbered 1, 2, ..., 600. Then one needs a

TABLE 1*

Ten thousand random digits

	00–04	05–09	10–14	15–19	20–24	25–29	30–34	35–39	40–44	45–49
00	88758	66605	33843	43623	62774	25517	09560	41880	85126	60755
01	35661	42832	16240	77410	20686	26656	59698	86241	13152	49187
02	26335	03771	46115	88133	40721	06787	95962	60841	91788	86386
03	60826	74718	56527	29508	91975	13695	25215	72237	06337	73439
04	95044	99896	13763	31764	93970	60987	14692	71039	34165	21297
05	83746	47694	06143	42741	38338	97694	69300	99864	19641	15083
06	27998	42562	63402	10056	81668	48744	08400	83124	19896	18805
07	82685	32323	74625	14510	85927	28017	80588	14756	54937	76379
08	18386	13862	10988	04197	18770	72757	71418	81133	69503	44037
09	21717	13141	22707	68165	58440	19187	08421	23872	03036	34208
10	18446	83052	31842	08634	11887	86070	08464	20565	74390	36541
11	66027	75177	47398	66423	70160	16232	67343	36205	50036	59411
12	51420	96779	54309	87456	78967	79638	68869	49062	02196	55109
13	27045	62626	73159	91149	96509	44204	92237	29969	49315	11804
14	13094	17725	14103	00067	68843	63565	93578	24756	10814	15185
15	92382	62518	17752	53163	63852	44840	02592	88572	03107	90169
16	16215	50809	49326	77232	90155	69955	93892	70445	00906	57002
17	09342	14528	64727	71403	84156	34083	35613	35670	10549	07468
18	38148	79001	03509	79424	39625	73315	18811	86230	99682	82896
19	23689	19997	72382	15247	80205	58090	43804	94548	82693	22799
20	25407	37726	73099	51057	68733	75768	77991	72641	95386	70138
21	25349	69456	19693	85568	93876	18661	69018	10332	83137	88257
22	02322	77491	56095	03055	37738	18216	81781	32245	84081	18436
23	15072	33261	99219	43307	39239	79712	94753	41450	30944	53912
24	27002	31036	85278	74547	84809	36252	09373	69471	15606	77209
25	66181	83316	40386	54316	29505	86032	34563	93204	72973	90760
26	09779	01822	45537	13128	51128	82703	75350	25179	86104	40638
27	10791	07706	87481	26107	24857	27805	42710	63471	08804	23455
28	74833	55767	31312	76611	67389	04691	39687	13596	88730	86850
29	17583	24038	83701	28570	63561	00098	60784	76098	84217	34997
30	45601	46977	39325	09286	41133	34031	94867	11849	75171	57682
31	60683	33112	65995	64203	18070	65437	13624	90896	80945	71987
32	29956	81169	18877	15296	94368	16317	34239	03643	66081	12242
33	91713	84235	75296	69875	62414	05197	66596	13083	46278	73498
34	85704	86588	82837	67822	95963	83021	90732	32661	64751	83903
35	17921	26111	35373	86494	48266	01888	65735	05315	79328	13367
36	13929	71341	80488	89827	48277	07229	71953	16128	65074	28782
37	03248	18880	21667	01311	61806	80201	47889	83052	31029	06023
38	50583	17972	12690	00452	93766	16414	01212	27964	02766	28786
39	10636	46975	09449	45986	34672	46916	63881	83117	53947	95218
40	43896	41278	42205	10425	66560	59967	90139	73563	29875	79033
41	76714	80963	74907	16890	15492	27489	06067	22287	19760	13056
42	22393	46719	02083	62428	45177	57562	49243	31748	64278	05731
43	70942	92042	22776	47761	13503	16037	30875	80754	47491	96012
44	92011	60326	86346	26738	01983	04186	41388	03848	78354	14964
45	66456	00126	45685	67607	70796	04889	98128	13599	93710	23974
46	96292	44348	20898	02227	76512	53185	03057	61375	10760	26889
47	19680	07146	53951	10935	23333	76233	13706	20502	60405	09745
48	67347	51442	24536	60151	05498	64678	87569	65066	17790	55413
49	95888	59255	06898	99137	50871	81625	42223	83303	48694	81953

*From Principles and procedures of statistics by R.G.D. Steel and J.H. Torrie. McGraw-Hill, 1960. Used with permission of McGraw-Hill Book Company, Inc.

TABLE 1 (Continued)

	50–54	55–59	60–64	65–69	70–74	75–79	80–84	85–89	90–94	95–99
00	70896	44520	64720	49898	78088	76740	47460	83150	78905	59870
01	56809	42909	25853	47624	29486	14196	75841	00393	42390	24847
02	66109	84775	07515	49949	61482	91836	48126	80778	21302	24975
03	18071	36263	14053	52526	44347	04923	68100	57805	19521	15345
04	98732	15120	91754	12657	74675	78500	01247	49719	47635	55514
05	36075	83967	22268	77971	31169	68584	21336	72541	66959	39708
06	04110	45061	78062	18911	27855	09419	56459	00695	70323	04538
07	75658	58509	24479	10202	13150	95946	55087	38398	18718	95561
08	87403	19142	27208	35149	34889	27003	14181	44813	17784	41036
09	00005	52142	65021	64438	69610	12154	98422	65320	79996	01935
10	43674	47103	48614	70823	78252	82403	93424	05236	54588	27757
11	68597	68874	35567	98463	99671	05634	81533	47406	17228	44455
12	91874	70208	06308	40719	02772	69589	79936	07514	44950	35190
13	73854	19470	53014	29375	62256	77488	74388	53949	49607	19816
14	65926	34117	55344	68155	38099	56009	03513	05926	35584	42328
15	40005	35246	49440	40295	44390	83043	26090	80201	02934	49260
16	46686	29890	14821	69783	34733	11803	64845	32065	14527	38702
17	02717	61518	39583	72863	50707	96115	07416	05041	36756	61065
18	17048	22281	35573	28944	96889	51823	57268	03866	27658	91950
19	75304	53248	42151	93928	17343	88322	28683	11252	10355	65175
20	97844	62947	62230	30500	92816	85232	27222	91701	11057	83257
21	07611	71163	82212	20653	21499	51496	40715	78952	33029	64207
22	47744	04603	44522	62783	39347	72310	41460	31052	40814	94297
23	54293	43576	88116	67416	34908	15238	40561	73940	56850	31078
24	67556	93979	73363	00300	11217	74405	18937	79000	68834	48307
25	86581	73041	95809	73986	49408	53316	90841	73808	53421	82315
26	28020	86282	83365	76600	11261	74354	20968	60770	12141	09539
27	42578	32471	37840	30872	75074	79027	57813	62831	54715	26693
28	47290	15997	86163	10571	81911	92124	92971	80860	41012	58666
29	24856	63911	13221	77028	06573	33667	30732	47280	12926	67276
30	16352	24836	60799	76281	83402	44709	78930	82969	84468	36910
31	89060	79852	97854	28324	39638	86936	06702	74304	39873	19496
32	07637	30412	04921	26471	09605	07355	20466	49793	40539	21077
33	37711	47786	37468	31963	16908	50283	80884	08252	72655	58926
34	82994	53232	58202	73318	62471	49650	15888	73370	98748	69181
35	31722	67288	12110	04776	15168	68862	92347	90789	66961	04162
36	93819	78050	19364	38037	25706	90879	05215	00260	14426	88207
37	65557	24496	04713	23688	26623	41356	47049	60676	72236	01214
38	88001	91382	05129	36041	10257	55558	89979	58061	28957	10701
39	96648	70303	18191	62404	26558	92804	15415	02865	52449	78509
40	04118	51573	59356	02426	35010	37104	98316	44602	96478	08433
41	19317	27753	39431	26996	04465	69695	61374	06317	42225	62025
42	37182	91221	17307	68507	85725	81898	22588	22241	80337	89033
43	82990	03607	29560	60413	59743	75000	03806	13741	79671	25416
44	97294	21991	11217	98087	79124	52275	31088	32085	23089	21498
45	86771	69504	13345	42544	59616	07867	78717	82840	74669	21515
46	26046	55559	12200	95106	56496	76662	44880	89457	84209	01332
47	39689	05999	92290	79024	70271	93352	90272	94495	26842	54477
48	83265	89573	01437	43786	52986	49041	17952	35035	88985	84671
49	15128	35791	11296	45319	06330	82027	90808	54351	43091	30387

TABLE 1 (Continued)

	00–04	05–09	10–14	15–19	20–24	25–29	30–34	35–39	40–44	45–49
50	54441	64681	93190	00993	62130	44484	46293	60717	50239	76319
51	08573	52937	84274	95106	89117	65849	41356	65549	78787	50442
52	81067	68052	14270	19718	88499	63303	13533	91882	51136	60828
53	39737	58891	75278	98046	52284	40164	72442	77824	72900	14886
54	34958	76090	08827	61623	31114	86952	83645	91786	29633	78294
55	61417	72424	92626	71952	69709	81259	58472	43409	84454	88648
56	99187	14149	57474	32268	85424	90378	34682	47606	89295	02420
57	13130	13064	36485	48133	35319	05720	76317	70953	50823	06793
58	65563	11831	82402	46929	91446	72037	17205	89600	59084	55718
59	28737	49502	06060	52100	43704	50839	22538	56768	83467	19313
60	50353	74022	59767	49927	45882	74099	18758	57510	58560	07050
61	65208	96466	29917	22862	69972	35178	32911	08172	06277	62795
62	21323	38148	26696	81741	25131	20087	67452	19670	35898	50636
63	67875	29831	59330	46570	69768	36671	01031	95995	68417	68665
64	82631	26260	86554	31881	70512	37899	38851	40568	54284	24056
65	91989	39633	59039	12526	37730	68848	71399	28513	69018	10289
66	12950	31418	93425	69756	34036	55097	97241	92480	49745	42461
67	00328	27427	95474	97217	05034	26676	49629	13594	50525	13485
68	63986	16698	82804	04524	39919	32381	67488	05223	89537	59490
69	55775	75005	57912	20977	35722	51931	89565	77579	93085	06467
70	24761	56877	56357	78809	40748	69727	56652	12462	40528	75269
71	43820	80926	26795	57553	28319	25376	51795	26123	51102	89853
72	66669	02880	02987	33615	54206	20013	75872	88678	17726	60640
73	49944	66725	19779	50416	42800	71733	82052	28504	15593	51799
74	71003	87598	61296	95019	21568	86134	66096	65403	47166	78638
75	52715	04593	69484	93411	38046	13000	04293	60830	03914	75357
76	21998	31729	89963	11573	49442	69467	40265	56066	36024	25705
77	58970	96827	18377	31564	23555	86338	79250	43168	96929	97732
78	67592	59149	42554	42719	13553	48560	81167	10747	92552	19867
79	18298	18429	09357	96436	11237	88039	81020	00428	75731	37779
80	88420	28841	42628	84647	59024	52032	31251	72017	43875	48320
81	07627	88424	23381	29680	14027	75905	27037	22113	77873	78711
82	37917	93581	04979	21041	95252	62450	05937	81670	44894	47262
83	14783	95119	68464	08726	74818	91700	05961	23554	74649	50540
84	05378	32640	64562	15303	13168	23189	88198	63617	58566	56047
85	19640	96709	22047	07825	40583	99500	39989	96593	32254	37158
86	20514	11081	51131	56469	33947	77703	35679	45774	06776	67062
87	96763	56249	81243	62416	84451	14696	38195	70435	45948	67690
88	49439	61075	31558	59740	52759	55323	95226	01385	20158	54054
89	16294	50548	71317	32168	86071	47314	65393	56367	46910	51269
90	31381	94301	79273	32843	05862	36211	93960	00671	67631	23952
91	98032	87203	03227	66021	99666	98368	39222	36056	81992	20121
92	40700	31826	94774	11366	81391	33602	69608	84119	93204	26825
93	68692	66849	29366	77540	14978	06508	10824	65416	23629	63029
94	19047	10784	19607	20296	31804	72984	60060	50353	23260	58909
95	82867	69266	50733	62630	00956	61500	89913	30049	82321	62367
96	26528	28928	52600	72997	80943	04084	86662	90025	14360	64867
97	51166	00607	49962	30724	81707	14548	25844	47336	57492	02207
98	97245	15440	55182	15368	85136	98869	33712	95152	50973	98658
99	54998	88830	95639	45104	72676	28220	82576	57381	34438	24565

SOURCE: Prepared by Fred Gruenberger, Numerical Analysis Laboratory, University of Wisconsin, Madison, Wis., 1952

TABLE 1 (Continued)

	50–54	55–59	60–64	65–69	70–74	75–79	80–84	85–89	90–94	95–99
50	58649	85086	16502	97541	76611	94229	34987	86718	87208	05426
51	97306	52449	55596	66739	36525	97563	29469	31235	79276	10831
52	09942	79344	78160	11015	55777	22047	57615	15717	86239	36578
53	83842	28631	74893	47911	92170	38181	30416	54860	44120	73031
54	73778	30395	20163	76111	13712	33449	99224	18206	51418	70006
55	88381	56550	47467	59663	61117	39716	32927	06168	06217	45477
56	31044	21404	15968	21357	30772	81482	38807	67231	84283	63552
57	00909	63837	91328	81106	11740	50193	86806	21931	18054	49601
58	69882	37028	41732	37425	80832	03320	20690	32653	90145	03029
59	26059	78324	22501	73825	16927	31545	15695	74216	98372	28547
60	38573	98078	38982	33078	93524	45606	53463	20391	81637	37269
61	70624	00063	81455	16924	12848	23801	55481	78978	26795	10553
62	49806	23976	05640	29804	38988	25024	76951	02341	63219	75864
63	05461	67523	48316	14613	08541	35231	38312	14969	67279	50502
64	76582	62153	53801	51219	30424	32599	49099	83959	68408	20147
65	16660	80470	75062	75588	24384	27874	20018	11428	32265	07692
66	60166	42424	97470	88451	81270	80070	72959	26220	59939	31127
67	28953	03272	31460	41691	57736	72052	22762	96323	27616	53123
68	47536	86439	95210	96386	38704	15484	07426	70675	06888	81203
69	73457	26657	36983	72410	30244	97711	25652	09373	66218	64077
70	11190	66193	66287	09116	48140	37669	02932	50799	17255	06181
71	57062	78964	44455	14036	36098	40773	11688	33150	07459	36127
72	99624	67254	67302	18991	97687	54099	94884	42283	63258	50651
73	97521	83669	85968	16135	30133	51312	17831	75016	80278	68953
74	40273	04838	13661	64757	17461	78085	60094	27010	80945	66439
75	57260	06176	49963	29760	69546	61336	39429	41985	18572	98128
76	03451	47098	63495	71227	79304	29753	99131	18419	71791	81515
77	62331	20492	15393	84270	24396	32962	21632	92965	38670	44923
78	32290	51079	06512	38806	93327	80086	19088	59887	98416	24918
79	28014	80428	92853	31333	32648	16734	43418	90124	15086	48444
80	18950	16091	29543	65817	07002	73115	94115	20271	50250	25061
81	17403	69503	01866	13049	07263	13039	83844	80143	39048	62654
82	27999	50489	66613	21843	71746	65868	16208	46781	93402	12323
83	87076	53174	12165	84495	47947	60706	64034	31635	65169	93070
84	89044	45974	14524	46906	26052	51851	84197	61694	57429	63395
85	98048	64400	24705	75711	36232	57624	41424	77366	52790	84705
86	09345	12956	49770	80311	32319	48238	16952	92088	51222	82865
87	07086	77628	76195	47584	62411	40397	71857	54823	26536	56792
88	93128	25657	46872	11206	06831	87944	97914	64670	45760	34353
89	85137	70964	29947	27795	25547	37682	96105	26848	09389	64326
90	32798	39024	13814	98546	46585	84108	74603	94812	73968	68766
91	62496	26371	89880	52078	47781	95260	83464	65942	91761	53727
92	62707	81825	40987	97656	89714	52177	23778	07482	91678	40128
93	05500	28982	86124	19554	80818	94935	61924	31828	79369	23507
94	79476	31445	59498	85132	24582	26024	24002	63718	79164	43556
95	10653	29954	97568	91541	33139	84525	72271	02546	64818	14381
96	30524	06495	00886	40666	68574	49574	19705	16429	90981	08103
97	69050	22019	74066	14500	14506	06423	38332	34191	82663	85323
98	27908	78802	63446	07674	98871	63831	72449	42705	26513	19883
99	64520	16618	47409	19574	78136	46047	01277	79146	95759	36781

random number consisting of three random digits in the chosen three consecutive columns, for each package to be sampled. Suppose the pencil hits row 48 and column 10, on the first page of Table 1. Then in row 48 take the three digits in columns 10, 11, and 12; these are 245. Then take package 245 for the sample. In row 49 in the same three columns one next finds 068. So, take package 68 for the sample. Then go to row 50 on the next double page and find 931 in columns 10, 11, and 12. This is outside the range of 1 to 600, so discard it and also the next number 842. Since the next one is 142, take package 142 for the sample. If it should ever happen that one comes to a number corresponding to a package already chosen, simply go to the next digit number, and so on, until the required number of sample packages has been assembled.

D FUNDAMENTAL PRINCIPLES

First, the number of sample units n refers to the number of units which are chosen separately and independently. For food in small packages, the sample unit is often the individual package. For bulk materials, it is necessary to imagine a number of sample units within each large container, of a size equal to that subsequently subjected to individual analysis. The concepts of random sampling, stratification and heterogeneity are relevant (see Section B, above). A sample of 1000 g consisting of ten units of 100 g individually collected at random would consist of 10 sample units ($n = 10$); whereas, if all 1000 g are taken from one place, n would equal 1; or, drawing two independent sample units of 500 g, $n = 2$.

Second, drawing a larger number of smaller sample units provides greater protection than drawing the same total weight of sample in fewer sample units, simply because there is a greater chance of hitting an unusually contaminated portion if the sample units are scattered.

Third, the effective sample of the population consists only of those sample units which are actually examined. If ten sample units are drawn, but only three are examined, the sampling plan is one in which $n = 3$, not $n = 10$, and the probability of accepting defective lots is correspondingly increased.

Fourth, the governing factor is not the proportion of the lot which we sample, but rather the total gross size of the random sample of the population (number and size of sample units) and the related acceptance/rejection criteria. In that case, a random sample of 100×5 g sample units does very nearly as well for a 500,000 g lot as it does for a lot of 10,000 g. The main reasons for drawing a somewhat smaller sample of the population (i.e. fewer sample units) from a small rather than from a large lot are (i) economics, and (ii) the greater importance of the decision on the large lot. Trying to sample a fixed proportion of the lot is not a sound concept.

Fifth is the concept of *Frame* vs *Consignment*. Whenever it is not possible to take a random sample from an entire consignment, but only from some accessible portion of the consignment, this portion of the consignment is called the 'frame.' Sample units are then chosen randomly from the frame. After analysis the sampling results and conclusions technically apply only to the frame, not the consignment.

E PRACTICAL CONSIDERATIONS

To use available resources to the best effect, all food cannot receive the same attention. The important foods, and lots, should receive the most time and effort, which means that they should be subjected to the most intensive sampling. What factors govern this decision?

(a) *Hazard* The most important factor is the hazard involved. How hazardous is the type (or types) of microorganism present; and how hazardous are the number(s) present? These questions are discussed more fully in Chapter 4. With increasing hazard, more or larger sample units are needed to minimize the probability of accepting a lot which should be rejected (i.e. the consumer's risk).

(b) *Uniformity* If there is good mixing of the food as a result of the manufacturing process (e.g. when the food is a mobile liquid), a relatively smaller number or size of sample units can be used. If, on the other hand, there is likely to be relatively heavy contamination in a few small places in the food, relatively larger numbers of sample units are needed.

(c) *Stratification* If it is known that there is 'stratification' within lots of the food (see Section B, above), a corresponding stratification can be used in selecting sample units.

(d) *Record of consistency* A consistently good record on a food from a specific supplier indicates that his processing is well done and reliable, and hence justifies reduced sampling or even omission of sampling on occasional lots. Discretion must be used in making this decision, with increasing confidence as the record accumulates.

(e) *Practical limitations* Since regulatory agencies never have the resources to test all imported lots, they must reduce the sampling plans to what is feasible. Most microbiological tests are laborious and slow; yet regulatory agencies cannot hold highly perishable foods pending results of analysis. Political or administrative pressures to reduce sampling will increase the probability of error.

In considering the above factors, it must be remembered that the ability to distinguish between good and bad lots often improves relatively little with increase in the number of sample units constituting the population sample. Indeed, in many instances reliability only goes up roughly as the square root of the size of the population sample, so that four times as large a sample will only halve the likelihood of making wrong decisions, when it might be advantageous to halve the value of n, in order to make another similar examination possible.

The size of the individual sample unit, however, reflects directly on reliability in a presence or absence situation (e.g., *Salmonella* testing). For a given sampling plan, to double the size of the sample unit will approximately halve the concentration of cells required to give the same frequency of discovery, assuming that the sensitivity of the method used is sufficient to detect the smaller ratio of test organism to others which may be present.

3

Appropriate sampling plans

This text is concerned primarily with plans that may be applied to lots presented for acceptance at ports or similar points of entry. Frequently, little or no information is available to the receiving agency about either the method by which the food was processed or the record of previous performance by the same processor. Consequently, sampling plans which depend on the nature of the frequency distribution are not suitable because the nature of the distribution is not known. Moreover, even if it were, the same product might be acceptably produced in a different way, leading to a different frequency distribution for which the particular recommendation might not be suitable. It is necessary, therefore, for the circumstances in question, to use sampling plans which do not depend on the nature of the frequency distribution. This chapter presents arguments for different types of plan, recommends two, and presents examples of each.

A ATTRIBUTE AND MEASUREMENT DATA

Acceptance or rejection of a lot could theoretically be based upon either an 'attribute' or a 'measurement.'

Attribute data represent 'go/no go' decisions about quality; in the present context, whether or not some particular organism occurs in numbers above a specified level (which may be zero). For instance, the number of 'positives' among a given number of sample units represents attribute data. When attribute data are used, the decision is based on the number of sample units 'positive,' i.e. giving results above or below the level specified. The probabilities afforded by attribute sampling plans are practically independent of distribution within the lot. This is an important practical advantage of attribute plans, because the distribution of the organism(s) in most consignments is largely unknown.

Measurements typically involve some continuous variable like concentration, e.g., the amount in parts per billion of some chemical residue in a sample unit of food. When measurement data are used as such, the decision to accept or reject a lot is commonly based on a summary statistic such as the average, e.g., the average standard plate count. Use of measurement data and of variables plans requires knowledge of the type of frequency distribution of the measurement, and there are few instances in which reliable information is available. Moreover, this text aims to provide generally applicable plans which can be used for products from different acceptable processes having different frequency distributions of the relevant quality index. Hence, variables plans are not considered appropriate for our purpose and are not treated here.

Measurement data can be converted easily to attribute data by using the numbers of sample units above and below the critical level, e.g., a specified plate count. Where bacteriological observations provide measurement data, it is suggested that they be converted to attribute data in this way, and handled by attributes plans. This is done throughout this book.

B DIVIDING A PRODUCT INTO CLASSES

A sample unit may be regarded as 'defective' if it contains any of certain dangerous microorganisms, or more than some chosen number of other microorganisms. The symbol m is used to represent the dividing line separating the sample units into the two classes: defective (values above m) and acceptable (values equal to or less than m). In the case of a dangerous microorganism, m is zero (though in practice this only means that it is not revealed by the method used).

Where microorganisms of some types (for example, indicator organisms) can be tolerated, microbiologists will usually be able to recognize three classes of quality, where any single sample unit may be wholly acceptable, marginally acceptable, or defective. The symbol m is then used to separate acceptable from marginally acceptable quality, while M is used to separate marginally acceptable quality from defective quality. This discussion on three classes of quality is resumed in Section D below.

C TWO-CLASS ATTRIBUTES PLANS

A simple way to make a decision to accept or reject a food lot is the following. The decision will be based on some microbiological test performed on several (n) sample units. This might be a test for the presence or absence of an organism (positive or negative); or it could be a plate count, to see whether or not the count is above some critical number, m. As explained in Chapter 1, the decision-making process is defined by two numbers. The first of these is the number of sample units (see Section D, page 16), represented by the letter n. If $n = 5$, one tests 5 sample units of material, that is, makes five tests. If $n = 15$, one tests 15 sample units, etc. The second number is the maximum allowable number of sample units yielding unsatisfactory test results: e.g., the presence of the organism, or a count above m. This acceptable number is given the letter c. Thus in a presence/absence type of decision on a food lot, the sampling plan $n = 10, c = 2$ means the following: take a sample of 10 sample units and make a test on each; then if 2 or fewer show the presence of the organism, accept the lot (in regard to this characteristic); but if 3 or more of the 10 show the presence of the organism, reject the lot (see Section H, page 8 for explanation of rejection).

The stringency of the sampling plan depends upon n and c, as already indicated in Chapter 1, Section F. The larger the value of n at a given c value, the better the food quality must be to have the same chance of being passed. Conversely, for a given sample size n, if c is increased the plan is more lenient and will more often pass food (i.e. there is a higher probability of acceptance, P_a) of a given quality whenever offered. Thus compared with $n = 10, c = 2$, the plans $n = 15, c = 2$, and $n = 10, c = 1$ are more stringent, while the plans $n = 5, c = 2$, and $n = 10, c = 3$ are more lenient. Probabilities of acceptance for a set of plans are given in Tables 2 and 3. Figure 2 displays the OC curves for a few of these plans, to illustrate the characteristics of various two-class attributes plans.

D THREE-CLASS ATTRIBUTES PLANS

This is a new type of plan (see Bray, Lyon, and Burr, 1973), devised for the situation just identified in Section B, where the quality of the product in terms of microbial criteria can be divided into three classes, and one chooses two levels of sample counts, m and M. A count above M for any sample unit is unacceptable. If there are any such counts among the n sample units from a lot, then this lot is withheld pending further investigation (see Section J, p. 54). Counts between m and M are undesirable, but some such counts can be accepted if there are not too many of them.

Accordingly, in 3-class plans there are again only two numbers, n and c, from which it is possible to find the probability of acceptance, P_a,

TABLE 2

Two-class plans ($c = 0$)

Probabilities of acceptance (P_a) of lots containing indicated proportions of acceptable and defective sample units

Composition of lot		Number of sample units tested from the population (n)							
% Acceptable (100-p)	% Defective (p)	3	5	10	15	20	30	60	100
98	2	0.94	0.90	0.82	0.74	0.67	0.55	0.30	0.13
95	5	0.86	0.77	0.60	0.46	0.36	0.21	0.05	0.01
90	10	0.73	0.59	0.35	0.21	0.12	0.04	<	<
80	20	0.51	0.33	0.11	0.04	0.01	<		
70	30	0.34	0.17	0.03	<	<			
60	40	0.22	0.08	0.01					
50	50	0.13	0.03	<					
40	60	0.06	0.01						
30	70	0.03	<						
20	80	0.01							
10	90	<*							

* '<' means $P_a < 0.005$

TABLE 3

Two-class plans (selected *c* values)

Probabilities of acceptance (P_a) of lots containing indicated proportions of acceptable and defective sample units													
Composition of lot		Number of sample units tested from the population (*n*) (top line) and *c* value (2nd Line)											
		5			10			15			20		
% Acceptable (100-*p*)	% Defective (*p*)	3	2	1	3	2	1	4	2	1	9	4	1
98	2	1.00	1.00	1.00	1.00	1.00	0.98	1.00	1.00	0.96	1.00	1.00	0.94
95	5	1.00	1.00	0.98	1.00	0.99	0.91	1.00	0.96	0.83	1.00	1.00	0.74
90	10	1.00	0.99	0.92	0.99	0.93	0.74	0.99	0.82	0.55	1.00	0.96	0.39
80	20	0.99	0.94	0.74	0.88	0.68	0.38	0.84	0.40	0.17	1.00	0.63	0.07
70	30	0.97	0.84	0.53	0.65	0.38	0.15	0.52	0.13	0.04	0.95	0.24	0.01
60	40	0.91	0.68	0.34	0.38	0.17	0.05	0.22	0.03	0.01	0.76	0.05	<
50	50	0.81	0.50	0.19	0.17	0.05	0.01	0.06	<	<	0.41	0.01	<
40	60	0.66	0.32	0.09	0.05	0.01	<	0.01			0.13	<	
30	70	0.47	0.16	0.03	0.01	<		<			0.02		
20	80	0.26	0.06	0.01	<						<		
10	90	0.08	0.01	<									
5	95	0.02	<*										

* '<' means $P_a < 0.005$

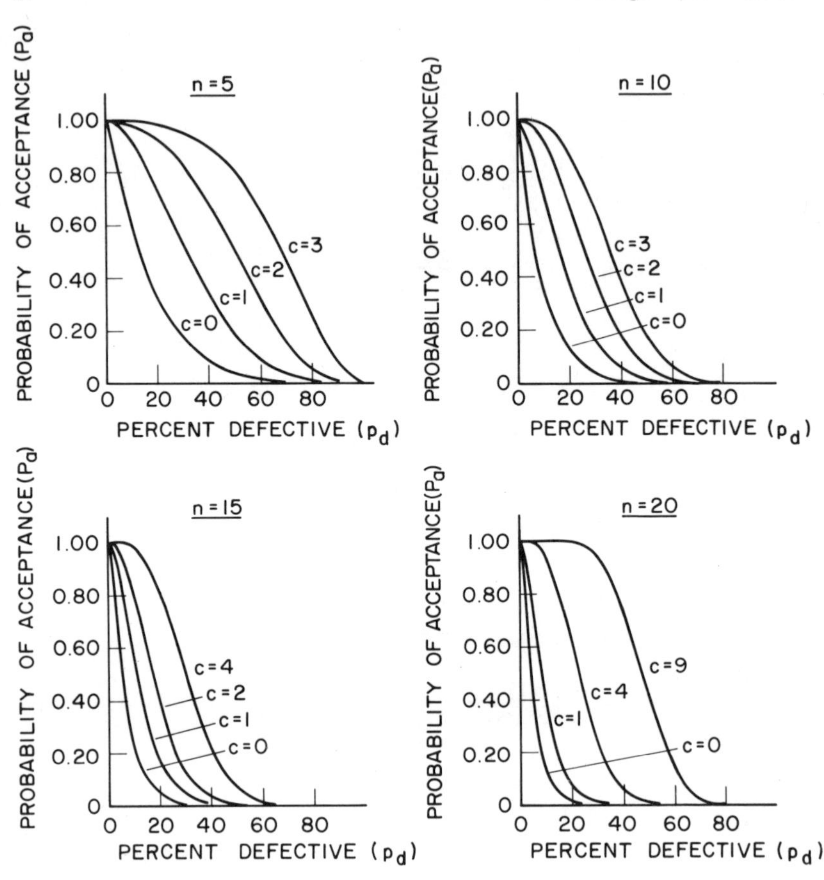

Figure 2 Operating characteristic curves for different sample sizes (n) and different criteria of acceptance (c) for two-class attributes plan

for a food lot of any given microbiological quality. To describe the lot quality we consider all sample units which could be drawn from the lot, which would yield counts in three classes: 0 to m, m to M, and above M. But, since the proportions in the lot for the three classes must add up to 1, we need only specify *two* of them in describing lot quality. We might call these proportions the per cent 'defective,' i.e. above M (P_d), and the per cent 'marginally acceptable,' i.e. m to M (P_m). Then the per cent 'acceptable,' i.e., 0 to m, is 100% minus the sum of the other two percentages. By appropriate calculations, we can then find the probability of acceptance, P_a, for a given lot quality for any specified sampling plan. For example, for the plan $n = 10$, $c = 2$, P_a will be 0.21 for a lot distribution for which 20% of the sample counts are 'marginally acceptable' and 10% are 'defective.' That is, on the basis of the particular values decided upon for m and M, only about 21 lots among 100 of that quality will be accepted, because they have no 'defective' counts and two or fewer 'marginally acceptable' counts out of the 10 samples chosen for the lot. The other lots will all be rejected if they have at least one 'defective' count and/or more than two 'marginally acceptable' counts among the 10 sample units.

Probabilities associated with a collection of 3-class plans are shown in Table 4 for various lot qualities.

E ATTRIBUTES PLANS AND CONSUMER PROTECTION

Table 5 compares the operating characteristics of the two types of plans recommended in this text, on the basis of equal sample sizes n, acceptance numbers c, and lot qualities. To facilitate the comparison, lot quality is measured as the proportion of the lot worse than level m and, correspondingly, the same value of c is taken for the 2-class plan as the c of marginally acceptable quality for the 3-class plan.

The 2-class plans do not distinguish values between m and M from those above M. If not more than c sample units give results above m, the lot is acceptable, regardless of how far individual results exceed m. But the corresponding 3-class plan does make a distinction. An extra subdivision of lot quality is made in the 3-class plan, since the limit M separates marginally acceptable from defective quality, and the difference between the two types of plan is the addition of the rejection criterion M. The same number of results exceeding m, i.e. value of c, is still permitted for acceptance. If none of them exceed M, all of these results will be acceptable, and the situation is as in the 2-class plan; thus the probabilities of acceptance are the same for the 2-class plan and the 3-class plan if there are no units in the lot (0%) yielding results above M. But if some of the results exceeding m also exceed M, the relevant sample units will be rejected on the basis of the M criterion. Therefore, in lots where an appreciable proportion (e.g., 5, 20, or 50%) of sample unit results exceed M, the probability of acceptance is correspondingly reduced. Thus the P_a values for the 3-class plans are all less than, or at best equal to, those for the corresponding 2-class plans. This is evident in Table 5.

Both consumer and producer protection is limited for population sample sizes n of 5 or 10. This can be (and in the proposed plans has

TABLE 4

Three-class plans: probabilities of acceptance (P_a) of lots containing indicated proportions for selected numbers of sample units tested and c values (p_d = per cent defective, p_m = per cent marginal)*

	p_m									
p_d	0	10	20	30	40	50	60	70	80	90
$n = 5$, $c = 3$										
70	<†	<	<							
60	0.01	0.01	0.01	<	<					
50	0.03	0.03	0.03	0.02	0.01	<				
40	0.08	0.08	0.07	0.06	0.04	0.02	<			
30	0.17	0.17	0.16	0.15	0.12	0.70	0.03	<		
20	0.33	0.33	0.32	0.30	0.27	0.20	0.12	0.04	<	
10	0.59	0.59	0.58	0.52	0.43	0.32	0.18	0.06	<	<
5	0.77	0.77	0.77	0.75	0.69	0.69	0.47	0.31	0.14	0.02
2	0.90	0.90	0.90	0.87	0.82	0.72	0.58	0.40	0.21	0.05
0	1.0	1.0	0.99	0.97	0.91	0.81	0.66	0.47	0.26	0.08
$n = 5$, $c = 2$										
70	<	<	<							
60	0.01	0.01	0.01	<						
50	0.03	0.03	0.02	0.01	<					
40	0.08	0.08	0.06	0.04	0.02	<				
30	0.17	0.16	0.14	0.11	0.06	0.02	<	<		
20	0.33	0.32	0.29	0.24	0.16	0.09	0.03	0.01	<	
10	0.59	0.58	0.55	0.47	0.36	0.23	0.12	0.05	0.01	<
5	0.77	0.77	0.72	0.63	0.50	0.35	0.20	0.09	0.02	<
2	0.90	0.90	0.85	0.75	0.60	0.43	0.27	0.13	0.04	<
0	1.0	0.99	0.94	0.84	0.68	0.50	0.32	0.16	0.06	0.01

TABLE 4 (Continued)

p_d	p_m 0	10	20	30	40	50	60	70	80	90
$n = 5, c = 1$										
70	<	<								
60	0.01	0.01	<							
50	0.03	0.02	0.01	<						
40	0.08	0.06	0.04	0.01	<					
30	0.17	0.14	0.09	0.05	0.02	<	<			
20	0.33	0.29	0.21	0.13	0.06	0.02	0.01	<		
10	0.59	0.53	0.41	0.27	0.16	0.07	0.03	0.01	<	
5	0.77	0.70	0.55	0.38	0.23	0.12	0.05	0.01	<	
2	0.90	0.83	0.66	0.47	0.29	0.16	0.07	0.02	<	
0	1.0	0.92	0.74	0.53	0.34	0.19	0.09	0.03	0.01	<
$n = 10, c = 3$										
50	<	<								
40	0.01	0.01	<	<						
30	0.03	0.03	0.02	0.01	<					
20	0.11	0.10	0.08	0.05	0.02	<	<			
10	0.35	0.34	0.29	0.20	0.10	0.03	0.01	<		
5	0.60	0.59	0.51	0.36	0.20	0.08	0.02	<		
2	0.82	0.81	0.71	0.52	0.30	0.13	0.04	0.01	<	
0	1.0	0.99	0.88	0.65	0.38	0.17	0.05	0.01	<	

TABLE 4 (Concluded)

p_d	p_m 0	10	20	30	40	50	60	70	80	90
$n = 10$, $c = 2$										
50	<									
40	0.01	<	<							
30	0.03	0.02	0.01	<	<					
20	0.11	0.09	0.06	0.02	0.01	<				
10	0.35	0.32	0.21	0.10	0.04	0.01	<			
5	0.60	0.55	0.39	0.20	0.08	0.02	<			
2	0.82	0.76	0.54	0.30	0.13	0.04	0.01	<		
0	1.0	0.93	0.68	0.38	0.17	0.05	0.01	<		
$n = 10$, $c = 1$										
50	<									
40	0.01	<								
30	0.03	0.02	<	<						
20	0.11	0.07	0.03	0.01	<					
10	0.35	0.24	0.11	0.04	0.01	<				
5	0.60	0.43	0.21	0.08	0.02	<				
2	0.82	0.59	0.30	0.11	0.03	0.01	<			
0	1.0	0.74	0.38	0.15	0.05	0.01	<			

* Each of these blocks of numbers, relating P_a to p_m and p_d, represents a three-dimensional relation called an OC surface, corresponding with the two-dimensional OC curve.

† '<' means $P_a < 0.005$

TABLE 5

Probabilities of acceptance (P_a) for two- and three-class attributes sampling plans when the lot examined contains 5%, 20%, or 50% of sample units > m

Sampling plan		% of sample unit results in the lot exceeding m										
		5%			20%				50%			
		2-class plan	3-class plan % > M		2-class plan	3-class plan % > M			2-class plan	3-class plan % > M		
n	c		0	5		0	10	20		0	25	50
5	0	0.77	0.77	0.77	0.33	0.33	0.33	0.33	0.03	0.03	0.03	0.03
5	1	0.98	0.98	0.77	0.74	0.74	0.53	0.33	0.19	0.19	0.11	0.03
5	2	1.00	1.00	0.77	0.94	0.94	0.58	0.33	0.50	0.50	0.19	0.03
5	3	1.00	1.00	0.77	0.99	0.99	0.59	0.33	0.81	0.81	0.23	0.03
10	0	0.60	0.60	0.60	0.11	0.11	0.11	0.11	<	<	<	<
10	1	0.91	0.91	0.60	0.38	0.38	0.24	0.11	0.01	0.01	0.01	<
10	2	0.99	0.99	0.60	0.68	0.68	0.32	0.11	0.05	0.05	0.02	<
10	3	1.00	1.00	0.60	0.88	0.88	0.34	0.11	0.17	0.17	0.03	<
15	0	0.46	0.46	0.46	0.04	0.04	0.04	0.04	<	<	<	<
20	0	0.36	0.36	0.36	0.01	0.01	0.01	0.01	<	<	<	<
60	0	0.05	0.05	0.05	<*	<	<	<	<	<	<	<

* '<' means $P_a < 0.005$

been) partially offset by choosing appropriate values for m and M such that, even if in a few instances lots are accepted with values exceeding the limits, the likelihood that the individual consumer will consume an entirely unsatisfactory product is remote. The choice of the appropriate values is based on experience; a project to lessen the subjective nature of such choice is described elsewhere in this text (see Conclusions, page 156). In addition, as the actual quality of a food worsens in relation to the stated limits, the probabilities of acceptance recede correspondingly. Hence, even a sampling plan with low numbers of sample units n (the extent of analysis generally feasible for economic reasons) can indicate a seriously substandard product, and the protection provided should suffice to exclude grossly inadequate products.

F RANDOM *VS* NON-RANDOM DISTRIBUTION WITHIN A LOT

Conditions of production, packing, and shipping may make the distribution of an organism far from random. Suppose, for example, that we have a lot of 2500 kg which may be considered to contain 100,000 25 g sample units, that we intend to use the sampling plan $n = 60$, $c = 0$, i.e., a total sample of $60 \times 25 = 1500$ g, and that there are 2500 *Salmonella* cells in the lot. They might be scattered at random throughout the lot, when there could be a maximum of 2500 units containing one cell each. Or, alternatively, the cells might conceivably be all in a few sample units, for example 100 containing an average 25 cells each.

The use of a 2-class attribute plan, with a test in which a sample unit is recorded as 'positive' however many *Salmonella* cells it may contain, is clearly likely to lead to different conclusions in the two alternative situations, as the proportion of defective sample units will approach 2500/100,000 ($p = 0.025$) with random distribution, but only 100/100,000 ($p = 0.001$) in the alternative example; and reference to the OC curve for the plan $n = 60$, $c = 0$ (given in Fig. 5, page 66) shows that for $p = 0.025$ (2.5%) the probability of acceptance P_a is only about 0.2, while for $p = 0.001$ it is about 0.95. The randomly contaminated lot would be accepted only about once in 5 (i.e., 4 out of 20) times, while the other would be accepted 19 out of 20 times.

As indicated in Figure 2 but more clearly illustrated in Figure 5, for plans with $c = 0$ the only way to decrease P_a for a given value of p is to increase n. Evidently, in using a 2-class plan for protection, one should choose a larger population sample (i.e. number of sample units n) when the distribution might be non-random. The probabilities afforded by

3-class plans are practically independent of distribution within the lot. This is an important practical advantage of this type of plan, because the distribution of the organism(s) in most consignments is largely unknown. If evidence can be accumulated indicating random distribution (good control) at a satisfactory level (low p), the population sample may be reduced for routine inspection.

G THE INFLUENCE OF SIZE OF LOT

A population sample of say $n = 30$ sample units might be taken at random from a lot of any size (or consignment if appropriate – see Section A, page 9) for use in a 2-class plan with $c = 0$. Suppose first that the lot contains a very large number of sample units. One then obtains an OC curve such as that shown on Figure 5 (page 66). There it is apparent that a lot with one defective unit out of 40 ($p = 0.025$) will be accepted about half the time ($P_a = 0.50$) when using the plan $n = 30$, $c = 0$. If the cells are spread more or less uniformly throughout the lot, this corresponds to a cell concentration of 1 in 40 × 25 g, i.e., one cell per kg, for sample units of 25 g. If, on the other hand, the lot contains a smaller number of sample units, the probabilities of acceptance P_a for various values of p would be only slightly smaller than those shown in Figure 5, despite the fact that the 30 sample units might now constitute an appreciable fraction of the whole lot. This improvement is quite small, until one begins to take a quarter to a half of the lot into the population sample, which could scarcely ever be done in bacteriological analysis of a food.

Suppose instead of $n = 30$, $c = 0$, we use $n = 5$, $c = 0$. The latter plan is quite undiscriminating. But the associated probabilities of acceptance (see Fig. 2, page 24) are again nearly independent of the lot (or consignment) size.

In summary, a sampling plan of the form described gives nearly the same degree of protection against acceptance of 'bad' lots or against rejection of 'good' ones, regardless of lot size.

4

Choice of a sampling plan according to purpose

A GENERAL PRINCIPLES

The stringency of sampling plans should be based on the hazard to the consumer from pathogenic or spoilage microorganisms, which is a function of the types of organisms present, and of their numbers. Some organisms merely spoil the product; some indicate the possibility of contamination by pathogens; some cause mild illness, and are unlikely to spread rapidly; some cause mild illness, but may spread rapidly; and some directly cause severe illness. The degree of hazard, of whatever type, is much increased if the organism grows to high numbers of cells in the food; and conversely, is usually reduced if the number is reduced. Treatment in the normal course of distribution, storage, and preparation for consumption, may decrease, leave unchanged, or increase the numbers.

The choice of a plan must therefore consider: (i) the seriousness of the type of hazard implied by the species of microorganism for which the test is to be made; and (ii) the future conditions to which the lot will usually be exposed, in relation to their likely effect on the number of cells and any corresponding change in degree of hazard. Table 6 classifies 15 different cases of sampling plans on a two-dimensional grid related to these two factors. The stringency of sampling increases with the type and degree of hazard: from a condition of no health hazard but only of utility, through a low indirect health risk (as implied by the presence of indicator species) to direct health risks related respectively to disease of moderate or severe implication. The most lenient plan is case 1. Stringency increases as one proceeds toward the right and toward the bottom of the table, so that case 15 is the most stringent.

On the basis of the above categorization into 'cases,' we now ex-

TABLE 6

Plan stringency (case) in relation to degree of health hazard and conditions of use

Type of hazard	Conditions in which food is expected to be handled and consumed after sampling, in the usual course of events		
	Reduce degree of hazard	Cause no change in hazard	May increase hazard
No direct health hazard			
Utility (e.g., general contamination, reduced shelf-life, and spoilage)	Case 1	Case 2	Case 3
Health Hazard			
Low, indirect (indicator)	Case 4	Case 5	Case 6
Moderate, direct, limited spread[a]	Case 7	Case 8	Case 9
Moderate, direct, potentially extensive spread[a]	Case 10	Case 11	Case 12
Severe, direct	Case 13	Case 14	Case 15

a See 'Conclusions,' p. 39, for explanation of extensive and limited spread

amine in more detail the considerations involved: in assessing the type of hazard associated with particular microbial species; in deciding whether likely conditions of use will affect the degree of hazard; in recognizing the corresponding case; and in choosing an appropriate sampling plan accordingly.

B MEDICAL AND EPIDEMIOLOGICAL FACTORS THAT INFLUENCE THE TYPE OF HAZARD

Among the many factors that determine the nature of the hazard from a species causing food-borne disease, and hence influence sampling, are frequency, clinical severity, duration, infectivity, the likelihood of a carrier state, and the extent to which the pathogen or its toxin may be distributed widely in the implicated or suspected food. Because clinical severity is of paramount concern, this factor should be given full weight when selecting the appropriate sampling plan. In general, the more severe the disease, the more stringent the plan required. In Table 7, the main food-borne pathogens are grouped in three categories, headed 'Severe Hazards,' 'Moderate Hazards, Potentially Extensive Spread,' and 'Moderate Hazards, Limited Spread,' according to the seriousness of the diseases they may cause (see Conclusions, page 39, for explanation

TABLE 7

Representative food-borne pathogens or toxins[a]

Organism	Frequency	Distribution	Vehicles[b]	Secondary factors that influence stringency of sampling
I SEVERE HAZARDS				
Clostridium botulinum (Botulism)	Rare in areas with effective food control	Widespread	Faultily processed (canned or preserved) foods; meat products; raw & smoked fish	High mortality; rapid recognition and specific treatments may be essential for patient survival
Salmonella typhi, *S. paratyphi* & *S. cholerae suis* (Typhoid & paratyphoid fevers)	Endemic in many parts of world; occasionally epidemic	Worldwide	Water; raw milk & milk products; meat products & vegetables	Low infective dose; require prolonged medical care; liable to result in carrier state (especially *S. typhi*)
Shigella dysenteriae I[c] (*Shigae*) (Shigellosis, Shiga dysentery)	Sporadic or epidemic	Central America, Mexico, North & Central Africa, Japan & Southeast Asia	Water, vegetable & salad foods	High mortality rate; frequently misdiagnosed
Vibrio comma (Cholera)	Sporadic, endemic & occasionally epidemic	Asia, Middle East, North & Central Africa	Water, various foods	
Brucella melitensis (Brucellosis)	Moderately rare, but occasionally endemic	Mediterranean countries	Goat milk & cheese	Convalescence often prolonged
Clostridium perfringens, type C (Enteritis necroticans)	Rare	Sporadic in Europe; New Guinea	Cooked meats	
Infectious hepatitis virus[d]	Common	Worldwide	Water, milk & milk products; salad foods; [illegible]	Very severe for patient with liver disease; long duration

TABLE 7 (Continued)

Representative food-borne pathogens or toxins[a]

Organism	Frequency	Distribution	Vehicles[b]	Secondary factors that influence stringency of sampling
II MODERATE HAZARDS: POTENTIALLY EXTENSIVE SPREAD[e]				
Salmonella typhimurium and other salmonellae spp. (Salmonellosis)	Common	Worldwide	Poultry & eggs; meats; wide range of other foods	Serious for the young & old
Shigella (see Section I) (Shigellosis, Flexner & Sonne dysentery)	Common: endemic in certain areas	Worldwide	Water, salads, fruits	Serious for young & old: difficult to isolate organism from foods
Vibrio parahaemolyticus	Common in Japan; increasingly reported from elsewhere	Far East Probably widespread	Marine fish; crustacea	Increased risk from raw or inadequately cooked fish
Escherichia coli (enteropathogenic)	Increasingly reported from several areas	Probably worldwide	Meats, raw milk & milk products	Serious for the young
Beta-haemolytic streptococcus	Infrequently food-borne	Widespread	Raw milk & milk products; egg salads	Certain group A types are liable to cause acute pharyngitis, nephritis, arthritis & cardiovascular complications

TABLE 7 (Concluded)

Representative food-borne pathogens or toxins[a]

Organism	Frequency	Distribution	Vehicles[b]	Secondary factors that influence stringency of sampling
III MODERATE HAZARDS: LIMITED SPREAD[e]				
Bacillus cereus	Increasingly reported from several areas	Probably worldwide	Reconstituted cereal products; milk, puddings & custards; rice	
Brucella abortus	Sporadic in several parts of world	Widespread	Raw milk & cream; fresh cheese	
Clostridium perfringens	Common	Probably worldwide	Cooked meats & poultry	
Staphylococcus (enterotoxigenic) (Staphylococcal entero-toxicosis or food poisoning)	Common	Worldwide	Ham, meat & poultry products; cream-filled pastries & exposed cooked foods; sauces & dressings; milk, cheese & egg dishes; crustacea	

a For fuller details of food-borne diseases, consult Bryan, 1971
b For diseases conveyed by milk, see also Joint FAO/WHO, 1970
c Sometimes *Shigella dysenteriae* causes mild food poisoning.
d The virus not yet isolated from foods
e See 'Conclusions,' p. 39, for explanation of extensive and limited spread

of extensive and limited spread). Sampling will be influenced, however, by other considerations, e.g., the frequency, geographic distribution, and characteristic vehicles of these diseases; the final column shows some secondary factors that affect the choice of sampling plan.

Table 7 refers only to the pathogens chiefly involved in bacterial food poisoning and other food-borne diseases, and is not intended to be exhaustive.[1] Omitted from the table, although occasionally implicated, are, for example, the 'Arizona group': *Pasteurella tularensis* (Ohara disease); *Mycobacterium tuberculosis*; *Listeria monocytogenes*; alpha-haemolytic streptococci; *Proteus* species; and *Pseudomonas aeruginosa*. Though anomalous, infectious hepatitis virus has been included, as a reminder that inability to isolate a bacterial pathogen from an implicated food neither assures the harmlessness of that food, nor necessarily indicates failure to employ adequate techniques and a sufficiently stringent sampling plan. No attempt has been made to arrange these pathogens according to the frequency with which they cause outbreaks or cases of food-borne disease, for the sequence would vary with locality. Botulism heads the list, despite its rarity, because of its unusually high mortality rate and the extraordinary precautions needed by processors of foods which may cause it.

Some basic medical and epidemiological considerations also affect the type of hazard, and thus the choice of plan. These will be briefly discussed under the headings aetiological, clinical, and epidemiological.

(a) *Aetiological*

Certain food-borne bacterial species are inherently associated with severe hazards for the human consumer. The 'virulence' of *Salmonella typhi*, *Vibrio comma*, and *Shigella dysenteriae* I (shigae) permit these organisms to survive and multiply in man, damage his health, and threaten his life. In these highly pathogenic species, the infective doses are apparently smaller than for less virulent organisms; but artificially induced immunity, or unfavourable environmental conditions near the portal of entry (e.g., gastric hyperacidity) may permit deceptively large inocula of otherwise virulent species to be ingested without harm.

1 In some areas of the world, protozoal or helminthic parasites are food-borne hazards even more important than bacterial types. For example, *Trichinella spiralis*, *Entamoeba histolytica*, and *Echinococcus granulosis* are severe hazards, and *Taenia solium*, *Taenia saginata*, and *Diphyllobothrium latum* are moderate hazards. These problems are outside the scope of this report; nevertheless, the same principles are applicable in such cases (see also Joint FAO/WHO Expert Committee on Zoonoses, 1959 and Dolman, 1957). The mycotoxins are likewise beyond the scope of this report.

Clostridium botulinum is regarded with apprehension because most strains can produce in many foods a toxin which proves fatal to man in very small amounts; yet, owing to uneven dispersal in the food, it is possible for routine samples to yield negative bacteriological and toxicological results. In such circumstances, the need for extensive 'investigational' examination in special situations is evident.

(b) *Clinical*

The clinical manifestations of a disease generally provide some clue to its severity. Hence the stringency of sampling plans may be affected by clinical considerations.

Shiga dysentery and botulism are examples of diseases with mortality rates which, without specific treatment, may reach 20 and 50 per cent respectively. In diagnosing these conditions, the laboratory often plays a crucial role. Botulism is a condition which few physicians are likely to have encountered; hence misdiagnoses are comparatively frequent, even when symptoms have been fairly typical. Further, for treatment of botulism to be effective, prompt diagnosis and type identification are essential. Again, the early differential diagnosis of Shiga dysentery from amoebic dysentery may depend mainly, if not solely, upon the laboratory. A shrewd clinician may suspect typhoid or paratyphoid fever, but the characteristic syndrome may not appear until several days after the laboratory could have identified the causal organism, thus permitting preventive action to be taken.

The relatively rare instances where beta-haemolytic streptococci are food-borne (usually by milk) may lead to tonsillitis complicated by severe sequelae such as glomerulonephritis, arthritis, and cardiovascular disabilities.

Moreover, in certain conditions, including the more moderate salmonelloses, shigelloses, and pathogenic *E. coli* enteritis, the severity of the infection becomes aggravated in the very young, the very old, or in individuals with pre-existing disabilities; even the usually moderate gastroenteritis caused by *V. parahaemolyticus*, staphylococcal enterotoxin, *B. cereus*, or *C. perfringens*, may sometimes become severe.

In some food-borne diseases convalescence may be lengthy, particularly after typhoid and paratyphoid fevers and brucellosis.

(c) *Epidemiological*

The propensity of many types of pathogenic bacteria to be distributed widely in the animal kingdom is another important factor that should tend to increase the stringency of sampling plans. In a food context, such rapid spread is characteristic of the genus *Salmonella*, for example.

The public health significance of the convalescent and chronic typhoid carrier is well established. The carrier state is less common, but still occurs, in the paratyphoid fevers, and may persist for months in other salmonelloses, as well as in shigellosis and pathogenic *E. coli* enteritis. The laboratory should be alert to this possibility, and plan its procedures accordingly.

In any part of the world, local customs and standards of community hygiene, especially those related to food, are important determinants of the extent and variety of food-borne infection. Prevailing standards for safeguarding water supplies, the quality of food-processing control, the fate of known contaminated or suspected foods, vermin control, public health supervision of restaurants, and intelligent use of refrigeration in processing plants, restaurants, and homes, all greatly affect the incidence of food-borne infection. Moreover, associations have been established epidemiologically between certain kinds of food-borne pathogens and specific vehicles: for example, *Salmonella typhi*, *Shigella dysenteriae* I (*shiga*), and *Vibrio comma*, and water; other types of salmonellae and poultry products and meats (although, of course, salmonellosis may be conveyed by many other foods); staphylococcal food poisoning and hams, cooked meats, poultry, and dairy products; *Bacillus cereus* and reconstituted dried cereal foods; *Vibrio parahaemolyticus* and marine fish; brucellosis and milk. Dietary customs also influence the food-borne hazards peculiar to a region. The Japanese liking for raw fish has favoured the relatively high incidence of Type E botulism and *V. parahaemolyticus* infection in that country. When the local regional existence of previously unrecognized kinds of food-borne infection has been established by the public health laboratory, its awareness soon changes to familiarity as other incidents are revealed. This interaction of circumstances no doubt helps to explain the recent rise in comparative frequency of enteritis due to *Clostridium perfringens* in the USA and to *B. cereus* in Britain.

(d) *Conclusions*

These considerations led the Commission to conclude that, among moderately hazardous species, it is justifiable to distinguish between those which present special epidemiological risks, and those which do not. The former, it is suggested, should include: *Salmonella*, *Shigella*, enteropathogenic *E. coli*, *Vibrio parahaemolyticus*, *Brucella*, and

β-haemolytic streptococci because, while they are primarily disseminated by specific foods, secondary spread to other foods commonly results from environmental contamination and 'cross-contamination' within factories and food preparation areas. Disease may result from relatively small inocula. The lower-risk group is represented by: *B. cereus*, *C. perfringens*, and enterotoxigenic *Staph. aureus* which are almost ubiquitous but cause disease only when the ingested foods contain very large numbers of the pathogens, or, in the instance of *Staphylococcus*, have at some time contained large enough numbers to have produced effective quantities of toxins. An outbreak is usually restricted to consumers of a particular meal. This distinction is represented in Tables 6 and 7, where the two groups are respectively referred to as causing 'potentially extensive' or 'limited' spread in foods. Less stringent procedures seem justified with organisms in the latter group.

Where the species represents a severe hazard, e.g., with *Shigella dysenteriae* type I or *C. botulinum* types A, B, or E, it seems undesirable to make such distinctions. Hence all such species are grouped together in Table 7.

C AN INDEX OF UTILITY

Utility tests, e.g., for general contamination, shelf-life, or spoilage, are not usually related to health hazard but rather to economics and aesthetics. Therefore, the corresponding cases 1–3 in Table 6 can be satisfied by relatively lenient plans. Generally speaking, they involve only standard plate counts, or specialized tests such as for total cold-tolerant organisms or, conceivably, species recognized as causing the relevant spoilage, e.g. lactobacilli. Such tests can, for example, confirm indications of spoilage based on odour, or predict the shelf-life of refrigerated foods.

D LOW HAZARD, INDICATOR TESTS

An inspector at the port of entry usually knows very little about the history of a consignment of food. He does not know whether a canning or cooking process was adequate to kill relevant bacteria, whether the food was contaminated after processing, or whether the shipment has been temperature-abused during shipment. Tests for indicator organisms frequently reveal related faults. High numbers of mesophilic spore-forming bacteria in non-acid shelf-stable canned goods indicate the likelihood of under-processing; enterococci, coliforms, and *E. coli* indicate inadequacy of general hygiene and sometimes demonstrate faecal contamination; staphylococci often show contamination from the human skin or nose (see also Thatcher and Clark, 1968).

Slightly more stringent plans than 'utility' tests are required for indicators of the likely presence of pathogens, even though the indication may be more or less uncertain. Foods should usually be tested for indicators rather than for pathogens, particularly when the pathogens are suspected to be present rarely or only in small numbers, or where feasible methods are not available for their routine detection (e.g., viruses).

E MODERATE AND SEVERE HAZARDS

One should apply tests for the pathogens themselves only under special circumstances: (a) If the history of sample analyses shows that the pathogen is frequently present and causes illness: for example, *C. perfringens* in cooked meat dishes; *B. cereus* in dried foods; *V. parahaemolyticus* in marine fish; salmonellae in red meats and poultry; and staphylococci in cooked foods, cheese, and fermented sausage. (b) When the epidemiology of a food-borne disease outbreak points to a particular lot of food as the cause of illness. (c) When there are other circumstances creating suspicion of the presence of the pathogen. For those instances in which the pathogens themselves are sought, the cases in Table 6 are further classified into 'moderate hazard' and 'severe hazard.' Table 7 has described these hazardous organisms in more detail.

In special circumstances, moderately hazardous organisms should be considered in the severe hazard category. For example, as shown in Table 8, very young, very old, or debilitated persons may become seriously ill or even die from salmonellosis. Persons on salt-free diets may become severely ill through loss of blood potassium simply because of diarrhoea.

The choice of plan to suit the difficult cases where special hazards exist is considered in Chapter 5; further illustration of factors influencing hazard in relation to sampling plans is given in Section B of that chapter.

F CHOOSING APPROPRIATE TESTS

A laboratory cannot possibly test a sample for all microbial groups, but must choose those that are appropriate to the problem in question. For such a choice, one must consider:

TABLE 8

Special foods for consumer groups with increased susceptibility

Food class	Reason for stringent sampling plan
Baby foods Geriatric foods	High susceptibility of the consumer population to enteric pathogens; severe response to toxins. Increased mortality
Low sugar food	Designed for diabetics for whom infection is a more severe risk than for others
Foods for hospitals	(a) Patients may be prone to infection and to serious sequelae after enterotoxicosis because of stresses from other disabilities and from immunodepressive treatment (b) Interference with convalescence from other disease (c) Staff and patients need to be protected because of their potential for disease spread within the hospital
Relief foods, especially dehydrated high protein foods distributed by international relief agencies	Consuming populations usually highly susceptible and prone to serious complications because of malnutrition and other stressful conditions. Increased risk for person-to-person spread of disease owing to confinement of the population and poor sanitary conditions. Reconstitution might permit rapid bacterial growth as storage conditions are usually inadequate

(a) *The food-borne disease record* of this type of food. For example, frozen eggs have had a history of causing *Salmonella* infections, but not botulism. Thus the plan for frozen eggs would include a test for *Salmonella* but not for *C. botulinum*.

(b) *The probable past history* of processing and storage. For example, one would ordinarily not test shelf-stable products in intact hermetically sealed cans for staphylococci, because effective retorting would have destroyed *Staphylococcus*. However, if ham receives a sufficiently low heat treatment, staphylococci could be of concern.

(c) *The expectation of subsequent spoilage*. For example, raw meats in chill storage will be spoiled by cold-tolerant bacteria, but do not present hazard from *C. botulinum*.

(d) *The age of the lot*. For example, a sample from a lot of canned foods processed six months earlier should not be incubated further to detect swells, whereas a sample from a newly processed lot might logically be incubated to determine whether it would be stable in further storage.

(e) *To what hazardous microorganisms the lot has been exposed*. For example, one might logically sample a fresh vegetable product for *Vibrio comma* if the product originated in the Orient, but not if it originated in Canada.

Such considerations will indicate the species of microorganisms likely to present hazards in the particular food under scrutiny. The choice of test methods, suitable for the various species likely to be involved, has been discussed in the preceding book (Thatcher and Clark, 1968).

The species and food being known, the considerations of the preceding sections of this chapter will indicate the associated type of hazard and the corresponding horizontal row in Table 6.

G THE EFFECT OF CONDITIONS OF USE ON DEGREE OF HAZARD

As noted in Section A of this chapter, hazard is related to the numbers of an undesirable organism in a food. Therefore, to decide which is the appropriate vertical column to use in Table 6, the effect of the conditions to which the food will probably be exposed in relation to the growth or death of the relevant microbes in the food must be considered.

If the food were ordinarily subjected to treatment that would permit bacterial growth, thereby increasing hazard, the case would be 3, 6, 9, 12, or 15: for example, clostridia surviving in cooked or pasteurized meats might grow in cans stored at ambient temperatures. If further treatment (e.g., freezing) would not change the number of relevant bacteria, one would choose cases 2, 5, 8, or 11: frozen desserts, such as cream pies, would be classified in one of these cases, because they would ordinarily be served immediately after thawing while still cold. If a food is expected to be fully cooked later, one would choose cases 1, 4, etc. since this condition reduces the hazard; raw meats would normally fall into this category.

For any particular lot of food, there will usually be one most appropriate vertical column in Table 6; but occasionally, different columns will apply to the different microbial species present, because of the differing characteristics of those species to grow, survive, or die in the food.

Among factors to be considered are recontamination, further storage conditions (i.e., frozen, ambient, hot warehouse), further processing (i.e., cooking, fermenting, washing, freezing), and consumption habits (i.e., eaten raw, warmed, cooked). The consideration can take into account only the usual treatment which the lot is expected to endure later. For example, a frozen food will ordinarily be kept frozen until thawed for cooking and/or serving; but if the inspector knows that in his country there is not enough freezer space, or that the handlers are usually careless, he may tighten the plan to take account of this expected abuse. If a food is unexpectedly abused after being sampled and passed, the conditions envisaged will, of course, not apply.

In deciding whether conditions of use will change hazard, the expected treatment must be considered in relation to the nature of the food in question, as it will affect each relevant bacterial species, because microorganisms differ in their behaviour. The following examples illustrate such considerations.

First, the nature of the food has to be considered in relation to the growth requirements of the relevant species. The infectious hepatitis virus will not grow at all outside the living cell; therefore, plans for this organism in foods would never allow for an increase in concern. Cooked cured ham frequently supports the growth of staphylococci which can grow at a water activity (a_w) as low as 0.86, whereas it rarely supports growth of salmonellae which do not grow at a_w below 0.94. Thus, cooked cured ham (if not refrigerated) might be classified in case 9 with respect to staphylococci, and in case 11 for salmonellae.

Meat gravy supports the growth of various bacteria, whereas dried beef does not: hence, if stored at ambient temperatures, hazard would increase in meat gravy corresponding to cases 3, 6, 9, 12, or 15; whereas with dried beef there would be no change in hazard from bacteria.

Potential growth may sometimes be prevented by the 'competitive' action of other microorganisms. While salmonellae grow in most foods of the necessary pH and a_w, staphylococci are often restricted by associated spoilage bacteria. Raw meats and bacon seldom cause staphylococcal poisoning, because they carry competitive species also, in large numbers; it usually arises from processed foods containing few organisms other than staphylococci – cooked ham for example. One should therefore regard staphylococci in raw meats or bacon as potentially case 7 or 8, whereas in cooked meats they are potentially case 9. Lack of competition occasionally allows toxin formation in foods not usually associated with staphylococcal food poisoning: for example, canned sardines contaminated by leakage, or cheese made from milk in which 'starter' activity is delayed. Microbial competition in food needs further investigation, for the outcome is often unpredictable.

Temperature relations are especially important. Microbial numbers, and associated hazards, generally increase at ambient temperatures and especially so at warm temperatures. On the other hand, refrigeration tends to reduce hazard, since most pathogens slowly die at low temperatures: for example, if ham were kept at temperatures below 6° C – a temperature at which staphylococci do not produce toxin, case 7 rather than case 9 would apply. The nature of the food and its associated species may be significant here. For example, storage of chilled smoked fish for long periods could permit toxin production by *C. botulinum* type E, because this organism can grow down to about 3° C. On the other hand, prolonged refrigeration of canned perishable ham creates no known hazard. Accordingly, with smoked fish one might choose case 15; with canned ham, case 5.

Consumption habits, also, affect hazard and the choice of 'case.' For example, *V. parahaemolyticus* grows readily on raw fish unless it is refrigerated. It is the major cause of food poisoning in Japan; but in North America, though widely distributed, it is a much less common cause of poisoning. The reason for this difference is that the Japanese eat raw fish, but North Americans cook it. Thus, in Japan one would choose case 12 for this species, whereas in the USA it would be case 10. Note that foods which are normally cooked (e.g., powdered eggs) may be consumed without cooking if they are distributed in relief foods. If a food is intended for consumers with unusually high susceptibility to food poisoning, hazard will thereby be increased (see Table 8, page 42).

When choosing cases on the basis of criteria of hazard, as above, one must consider whether the consignment under examination will be used exclusively for a single purpose. A consignment might include lots to be used for a variety of purposes. If so, it needs to meet a standard providing acceptable protection against the most 'sensitive' use to which its components will normally be put. An exception can be made if it is possible to treat the component lots individually, each according to its controlled use. In this eventuality, the individual acceptance criteria might be permitted to deviate from an overall standard.

H CHOOSING THE 'CASE'

In general, foods which have received an anti-bacterial heat treatment are safer. Hazard is likely when foods (i) become recontaminated after processing, (ii) are used in such a way as to permit multiplication of pathogens, (iii) are not cooked shortly before consumption. These principles must always be considered when estimating the degree of hazard, to decide case and sampling plan. Table 9 is intended to illustrate in more detail, with a single food (dried whole egg), the considerations involved in choosing the right 'case,' according to the various organisms or toxins which could be of concern. Note that the sources and degree of hazard between non-pasteurized and pasteurized egg are different; and that the discussion would not apply to dried egg albumin, because it has different properties (e.g., it is much less rich nutritionally, and more rich in anti-bacterial factors, especially for Gram-positive bacteria).

Foods derived from dried whole egg are used in diverse ways. They may be cooked and eaten at once (e.g., scrambled egg, custard), or cooked into products which thereafter prevent multiplication (e.g., cake or pasta); both these treatments diminish microbiological hazards. They may be eaten dry (e.g., dusting powder), or reconstituted and eaten promptly; these treatments do not change hazard. Sometimes, there may be long delays between reconstitution and consumption, when hazards increase. The cases are set out accordingly in Table 9.

As regards utility, a Standard Plate Count (SPC) would provide a useful index with the unpasteurized product. But with the pasteurized product, SPC more nearly reflects the result of the process and an offensively spoiled raw material might meet normal SPC criteria after pasteurization. Here a Direct Microscopic Count (DMC) is a desirable additional test for the number of bacteria present before pasteurization. With DMC or SPC, the case – 1, 2 or 3 – depends simply on subsequent use, as in Table 9.

Choice of case with the indicators – coliforms, faecal coliforms, or *E. coli* – is aided by the following considerations. Unpasteurized egg usually contains various species of Enterobacteriaceae in large numbers. The coliforms indicate general contamination, from dirty egg shells, cracked eggs, or unsanitary egg-breaking and pulp-mixing operations; large numbers indicate fault in one or more of these. Some tolerance is, however, necessary with the unpasteurized product because, from the nature of the materials and the operations, it cannot be produced without coliforms. Consequently this product should be used in foods which are heat processed (case 4). In the pasteurized product, on the other hand, coliforms will have been destroyed; any present will usually be derived from post-processing contamination, and tolerance can be low. If either product is to be used in a food that will be cooked (case 4), differentiation within the coliform group is unnecessary because all Gram-negative bacteria will be destroyed during cooking, and general level of contamination is the only concern. If either product undergoes delay after reconstitution, the case is 6, and differentiation of faecal coliforms is preferable because this group gives more certain indication of contamination from a faecal source. Confirmation of *E. coli* is in any case not warranted, because the number of *E. coli* gives little more information than the number of faecal coliforms for which much simpler tests suffice.

TABLE 9

Test organisms in relation to food use, treatment and case for dried whole-egg products,[a] pasteurized and unpasteurized (applicable, after selection, to routine or investigational analyses)

Type of hazard / Degree of hazard	Reduced		Not changed		Increased	
	Used in a food to be cooked, e.g., cake-mix, pasta		Consumed dry (e.g., dusting powder), or immediately after reconstitution		Delay after reconstitution and no cooking before consumption	
	Unpasteurized	Pasteurized	Unpasteurized	Pasteurized	Unpasteurized	Pasteurized
Utility	Case 1 SPC	Case 1 SPC DMC	Case 2 SPC	Case 2 SPC DMC	Case 3 SPC	Case 3 SPC DMC
Low indirect (Indicator)	Case 4 Coliforms	Case 4 Coliforms	Case 5 Coliforms	Case 5 Coliforms	Case 6 Faecal coliforms	Case 6 Faecal coliforms
Moderate, limited spread	Case 7 *Staphylococcus*: lenient tolerance; enterotoxin test before rejection	Case 7 —	Case 8 *Staphylococcus*: low tolerance enterotoxin test before rejection	Case 8 *Staphylococcus*: low tolerance	Case 9 *Staphylococcus*: minimal tolerance *C. perfringens* *B. cereus*	Case 9 *Staphylococcus*: minimal tolerance *C. perfringens* *B. cereus* Enterococci
Moderate, extensive spread	Case 10 *Salmonella* test to check recontamination risk	Case 10 —	Case 11 *Salmonella*	Case 11 —	Case 12 *Salmonella*	Case 12 *Salmonella* (recontamination)
Severe	Case 13 —	Case 13 —	Case 14 —	Case 14 —	Case 15 —	Case 15 *C. botulinum* where extreme abuse is suspected after recon-stitution

a Several of these items are not appropriate for dried white of egg.

With the moderate hazard provided by coagulase-positive *Staphylococcus aureus* (i.e., probably enterotoxin-forming), one could be lenient towards dried unpasteurized whole-egg used as a food to be cooked (case 7). A few such staphylococci would probably be introduced when breaking the eggs, but would be of little concern unless conditions before drying allowed production of enterotoxin. Because staphylococci tend to diminish in number in a dried food, it would be prudent to test for enterotoxin before rejecting a lot, if means were available. On the other hand, for the pasteurized product a low tolerance would be logical for staphylococci, because there they indicate post-processing contamination. This hazard would seem negligible if the product were to be cooked (case 7) as well as pasteurized; but it could become substantial under conditions of abuse (case 9). Staphylococci could be more hazardous in the pasteurized product because of reduced competition from other organisms. Whether or not the egg was pasteurized, minimal tolerance for staphylococci is indicated when there is reconstitution and delay before consumption, because of the probability of multiplication and toxin production (case 9).

Pasteurization would leave a residue of spores, including perhaps the moderate hazards *B. cereus* or *C. perfringens*, and enterococci. This would be unimportant if the whole-egg product were used in the dry state, or in a food that was cooked. But it could be important if there was a delay between reconstitution and consumption without cooking (case 9); with prolonged delay, *C. botulinum* could become a problem (case 15). In each instance, the critical factor would be how long the reconstituted product remained at temperatures favouring growth.

The moderately hazardous salmonellas often occur in egg pulp, and drying cannot be relied upon to destroy them or other Enterobacteriae. Consequently, it is advisable to test for *Salmonella* in dried unpasteurized whole egg: if used dry, there is no change in hazard (case 11); if reconstitution were delayed, and heating did not take place before consumption (highly undesirable with this product!), the case would be 12. Cooking promptly after reconstitution would remove the hazard, but it might be useful to check for the smaller hazard of recontamination (case 10). The pasteurized product should be free of salmonellas. Recontamination is a possibility, but the level would probably be negligible unless circumstances favoured multiplication. There should, therefore, be no need to test the pasteurized product for *Salmonella*, unless prospective conditions of use increase hazard. Then the case would be 12 (Table 9). Similar considerations would apply to *Shigella*.

The foregoing illustrates and emphasizes the need for some knowledge of the microbial ecology of a food before an examiner can choose an appropriate case and test for a particular purpose. Applying the principles outlined, Table 10 gives numerous examples of the relations between cases and conditions of handling and usage, and the organisms of concern, for various foods illustrating the classes considered in Part II.

I DECIDING BETWEEN TWO-CLASS OR THREE-CLASS PLANS

To decide whether the plan should be 2-class or 3-class, consider whether one can permit any positives (e.g., *Salmonella* or SPC levels above those reflected by good manufacturing practice) in any of the sample units. If the answer is no, use a 2-class, plan with $c = 0$. If the answer is yes, a 2- or 3-class plan can apply, but a 3-class plan is recommended (Fig. 3).

Three-class sampling plans, where applicable, have properties that should encourage their adoption in microbiological examination, for the following reasons:

(i) To accept a proportion of sample units yielding test values in the marginally acceptable interval (between acceptable and defective) is in keeping with practical experience where, even under good manufacturing conditions, a few specimens may well reveal test values beyond those normally encountered without causing any consequent problem. This situation applies especially to counts of 'indicator' organisms.

(ii) Sufficient experience also defines a different level, beyond which counts indicate substantial likelihood of health or utility hazard, a level which will not be attained if control has been adequate. This, designated M, should remain stable unless new experience reveals an error in its initial placement. Obviously, greater stability of criteria of acceptance may be expected to promote wider adoption of sampling plans.

(iii) As noted earlier (page 31), 3-class plans are less affected by unknown changes in the distribution of organisms within lots.

(iv) A 3-class plan allows a regulatory authority to warn a manufacturer of the desirability of in-plant testing, if sampling reveals a trend to increasing frequency of values within the marginally acceptable (m to M) range. Such warning can help the manufacturer reduce the likelihood of need for more drastic action in the future. It is therefore recommended that 3-class plans be used whenever count or concentration tests are performed.

TABLE 10

Relationship of 'cases' (degree of concern)[a] to food type, microbiological test and conditions of food treatment – illustrative examples from various classes of foods

Nature of concern	Effect of conditions	Case	Food product	Bacterial test	Anticipated conditions of treatment
Spoilage and shelf-life	Reduce hazard	1	Fresh and frozen fish	SPC	Cooked prior to consumption
			Frozen eggs	SPC	Used as an ingredient in a heat-processed mixture (e.g., cake)
			Frozen prepared foods	SPC	Eaten immediately after thawing and effective cooking
			Dried powdered foods (milk, relief foods)	SPC	Consumed immediately after reconstitution and heating
	No change in hazard	2	Cold smoked fish	SPC	Eaten without cooking
			Dried fruits	Moulds, yeasts	Well-packaged; storage under condition preventing water uptake
			Frozen egg whites	SPC	Frozen storage, further processing; promptly consumed without cooking
			Frozen cooked shrimp and lobster tails	SPC	Stored frozen, eaten upon thawing
	Increase hazard	3	Powdered dried foods	SPC	Poor packaging or in bulk; storage in humid conditions
			Nuts, grains, dried fruits	Moulds	Storage in humid conditions
			Raw meats	SPC	Prolonged storage; inadequate or discontinuous refrigeration
			Canned cured shelf-stable meats	'Swells'	Storage at normal ambient temperatures
			Canned vegetables and mixtures	Thermophiles	Storage at high temperatures
			Frozen foods	SPC	Uncertain adequacy of frozen storage, or prolonged thawing

TABLE 10 (Continued)

Nature of concern	Effect of conditions	Case	Food product	Bacterial test	Anticipated conditions of treatment
Low indirect health hazard (indicator organisms)	Reduce hazard	4	Cold smoked haddock	Faecal coliforms Staphylococci[b]	To be cooked
			Blanched, frozen vegetables	SPC Coliforms	To be cooked
			Cheese	Faecal coliforms	To be converted to processed cheese
			Dried foods	Coliforms	Used in a mixture which will be heated
			Frozen raw shrimp	Coliforms Staphylococci[b]	Cooked or heat-processed before consumption
	No change in hazard	5	Fresh or chilled vegetables	Coliforms	Eaten raw
			Cold smoked fish	Faecal coliforms	Frequently stored at room temperature
			Hand-picked cooked crab meat	Coliforms Staphylococci[b]	Not cooked before eating
			Dried milk[c]	SPC Coliforms	Consumed promptly after reconstitution
			Frozen vegetables or sauces	SPC Coliforms	Eaten without effective cooking
			Frozen cooked shrimp	Coliform Staphylococci[b]	Eaten immediately after thawing
			Frozen pre-cooked dinners or desserts	SPC Coliforms	Eaten without effective cooking
	Increase hazard	6	Raw or frozen shellfish	Coliforms	Derived from polluted or unknown water perhaps containing salmonellae, etc.
			Dried milk, relief foods	SPC Coliforms Staphylococci[b]	Anticipated conditions of poor control after reconstitution
			Frozen foods in general	SPC Coliforms Staphylococci[b]	In countries lacking sufficient effective freezer capacity

TABLE 10 (Continued)

Nature of concern	Effect of conditions	Case	Food product	Bacterial test	Anticipated conditions of treatment
Moderate direct health hazard, limited spread	Reduce hazard	7	Dried meats or components	Staphylococci[d]	To be heated after reconstitution
			Frozen cooked dinners	Staphylococci[d]	To be effectively reheated
	No change in hazard	8	Fresh chilled or frozen salad vegetables	*Salmonella*	To be eaten raw
			Cold smoked fish	Staphylococci[d]	Frequently stored at room temperature
			Dried foods, dried milk	Staphylococci[d]	To be consumed immediately after reconstitution
			Ice cream	Staphylococci[d]	Eaten frozen
			Cheese made from pasteurized milk	Staphylococci[d]	Normal storage
			Frozen desserts	*Salmonella*	To be eaten without thawing
	Increase hazard	9	Cooked meat	*C. perfringens*	Delay in cooling
			Ready cooked starchy foods, mayonnaise	*B. cereus*	Delay in consumption
			Cheese	Staphylococci[d]	Addition to other foods
			Dried milk, relief foods	Staphylococci[d]	Under conditions of inadequate control after reconstitution
Moderate direct health hazard, of potentially extensive spread in food	Reduce hazard	10	Raw meats & poultry	*Salmonella*	Cooked
			Fresh vegetables	*Salmonella* *Shigella*	Cooked
			Egg products	*Salmonella*	Cooked
	No change in hazard	11	Fresh and frozen fish	*V. parahaemolyticus*	In countries where eaten fresh raw
			Fresh-water fish from warm waters	*Salmonella*	Refrigerated
			Frozen cooked shellfish	*V. parahaemolyticus*	Eaten directly after thawing

TABLE 10 (Concluded)

Nature of concern	Effect of conditions	Case	Food product	Bacterial test	Anticipated conditions of treatment
	Increase hazard	12	Raw meats and poultry	*Salmonella*	Inadequate refrigeration, and cross-contamination
			Fresh and frozen fish	*V. parahaemolyticus*	Used for fermented products, eaten uncooked
Severe direct health hazard	Reduce hazard	13	Perishable meats	*C. botulinum*	Refrigeration
			Poultry meat	*Salmonella*	When used for sensitive populations (e.g., hospitals): with cooking
			Powdered egg		
			Raw vegetables	*V. comma*, *Sh. dysenteriae I*	Derived from endemic or epidemic areas, but cooked before eating
	No change in hazard	14	Shellfish	*S. typhi*	Derived from endemic or epidemic areas or from polluted waters, eaten raw
			Cheese made from pasteurized milk	Staphylococcal enterotoxin	Use by sensitive populations
	Increase hazard	15	Smoked fish	*C. botulinum* E	Normal refrigeration may permit toxin formation
			Perishable (non-shelf-stable) canned meats such as cured meats	*C. botulinum* B	Held under warm conditions; insufficient salt or nitrite; stored normally
			Canned non-acid foods	*C. botulinum* A & B	Foods from plants with inadequate process control

a Compare with Table 6, p. 33.
b *Staphylococcus aureus* (used as an indicator of insanitation, not as a direct health hazard)
c See also case 6.
d The concern here is the production of staphylococcal enterotoxin.

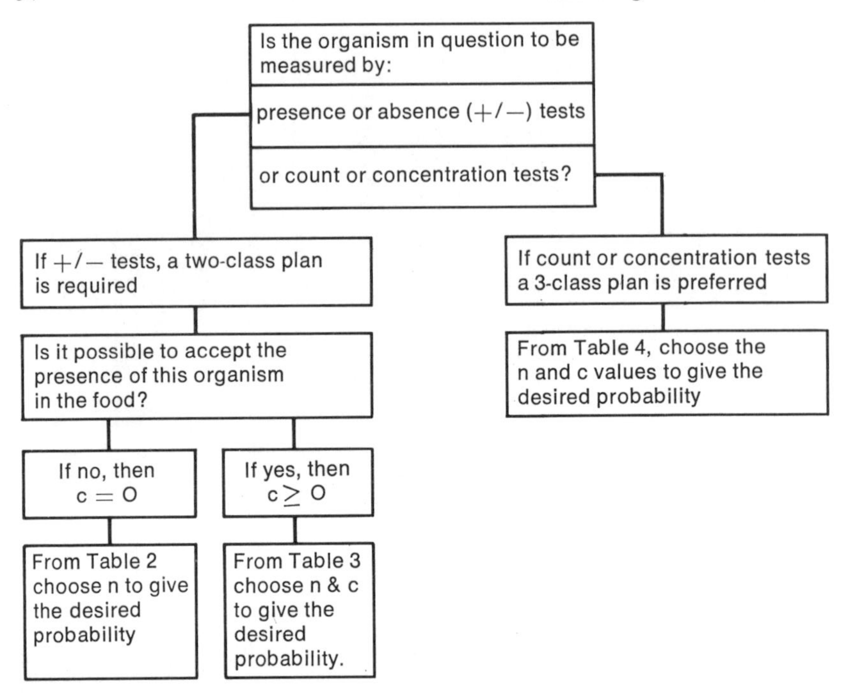

Figure 3 Choosing sampling plan for a specific application

J DETERMINING VALUES FOR m AND M

Preliminary definitions of m and M have been given in Section B, page 20.

The level of the test organism which is acceptable and attainable in the food is m. It is reflected by good commercial practice (GCP) as recognized either in the exporting country or, more often, in the domestic production of the importing country. As regards presence or absence of pathogens (i.e., for 2-class plans) m is 'zero'; or more correctly, a small number corresponding to the level of detectability in the test (see Section B, page 67). Hence, in this book, the value of m for 2-class plans is usually 0. For 3-class plans m will usually be assigned some non-zero value.

M, used only for 3-class plans, is a hazardous level of contamination caused by bad sanitary practice (including improper storage). All lots with values exceeding M should be rejected (in the sense defined in Section H, page 8; that is, withheld pending further investigation to determine whether the product can be used after further treatment or whether it is unacceptable as food). When such lots are detected, investigation of the producer's facilities should be undertaken immediately. In international commerce this presupposes a system whereby the analytical results can be conveyed to appropriate officials of the country of origin; but present systems are largely *ad hoc* or non-existent, except for shellfish (Chapter 16).

There are several approaches to decide the value of M:

(i) As a utility (spoilage or shelf-life) index. Relate levels of bacteria to detectable spoilage (odour, flavour) or to a decrease in shelf-life to an unacceptably short period.

(ii) As a general sanitation indicator. Relate levels of the indicator bacteria to a clearly unacceptable condition of sanitation – whether contamination or growth, or a combination of these factors.

(iii) As a health hazard. Relate levels of bacteria to illness. Use epidemiological and laboratory data in combination, experimental animal feeding or inoculation, human feeding data, laboratory analyses for toxin related to levels of organisms, or other guides that show the level at which definite hazard exists. For this purpose, consider the maximum amount of food likely to be eaten at one time and the group of persons likely to eat the food (see also Table 8, page 42).

This usage of m and M differs from that described by Kampelmacher *et al.* (1969), where the difference between m and M was thought of as an arbitrary fixed ratio. Such an arbitrary ratio has sometimes been used as a so-called administrative means of accommodating to limitations of reproducibility in an analytical method, or to the use of too few specimens for statistically valid decision. For example, a limit for the Standard Plate Count might be SPC 10^4/g and, because of inability to examine sufficient sample units, the product would be considered defective only when the found average exceeded, say, three times the stated limit. Such practice uses a principle quite different from that employed in the 3-class plan. Adherence to a specific sampling plan is much to be preferred.

All probability computations (e.g., OC curves and surfaces) for sampling plans assume that the laboratory result is obtained without error. The decision to accept or reject any particular lot will, of course, be based upon the particular laboratory results obtained, and is subject to the error associated with the laboratory procedure used. It is especially important when using 2-class plans with $m = 0$ and $c = 0$ that the methods employed give accurate answers, for whatever false negatives and false positives occur will result in corresponding wrong decisions about lots because of the inaccurate methods. When m and/or M are not

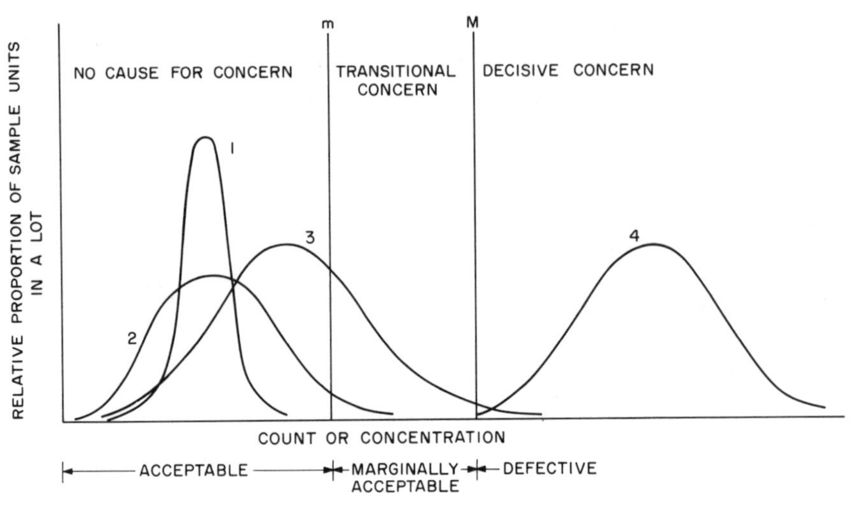

Figure 4 Considerations in choosing *m* and *M*

zero, it is recommended that these values be chosen in order to include the error of the method within the acceptable region for *m* and the marginally acceptable region for *M*. In this way acceptance of bad lots resulting from the error of the method can be avoided. This increases the producer's risk and favours the consumer, to the extent of the variation of the method. It follows, therefore, that the method used for enumeration influences the choice of the criteria *m* and *M*, and hence the effect of a given plan. Unfortunately, documentation of the inherent variation within a given method is rare, though experienced analytical bacteriologists develop estimates which become included in their choice of criteria. Some methods, for example MPN methods for coliforms, have such inherent variability that criteria based on them should not be stringent. Similarly, it is commonly found that long familiarity of an analyst with a particular method (e.g., for *Salmonella* or *Staphylococcus*) allows that method to be most productive *for him*. This is a cogent argument for widely accepted use of a standard method so that all analysts become experienced in its use: another method with different sensitivity might require change of criteria. Because the testing method is so important in deciding the acceptance criteria of sampling plans, it seems desirable not to change a standard method unless very real advantage in so doing has been demonstrated.

Figure 4 illustrates the effect of various frequency distributions of microbial content, within a lot, on the location of *m* and *M* for 3-class plans. Curve 1 represents an entirely satisfactory lot, with small numbers of bacteria generally, and thus a low average count with little variation, so that no counts exceed m. Curve 2 represents a lot with a similar average count, but with a much wider variation, so that a small proportion of sample units would have counts exceeding m, though none exceed M: if the proportion in the range m–M were small, the situation would be acceptable; if this proportion were larger it might still be acceptable, but it would call for a warning to the producer, as tending toward the situation shown in curve 3. Curve 3 represents a lot with a higher average count and large variation, such that a small proportion of sample units exceed M and would cause immediate rejection, while a substantial proportion fall in the range m–M, which itself could also suffice to justify rejection (see Section H, page 8 for explanation of the term 'rejection'). Curve 4 represents a lot of even greater unacceptability, requiring rejection. In an unusual situation, one might find a consignment composed of two widely dissimilar lots – one entirely acceptable (curve 1) and the other entirely unacceptable (curve 4); this situation would show a high proportion of units falling below m and above M, but few between m and M.

These examples illustrate how the choice of m and M, in relation to the microbial quality of the lot, influences the proportions of material likely to be rejected. Moreover they illustrate the importance to producers of limiting variation in numbers of microbes present in the food, i.e. of keeping production 'under control' in the bacteriological sense, to avoid lots of otherwise satisfactory average quality being rejected. With good or bad control, the average microbial levels are likely to be respectively low or high, and the normal distribution of data above the average (expressed logarithmically) respectively narrow or broad. The spread of the distribution may be nearly as important as the average level in determining compliance or non-compliance with a particular criterion, as Figure 4 illustrates.

K SPECIFIC KNOWLEDGE ABOUT THE LOT

The relatively long delay between collection of samples at a port of entry and issue of the corresponding laboratory result may create a need for costly storage. For products with an extensive history of good quality, the need can be avoided by releasing consignments as soon as samples have been taken (provided, of course, that records are adequate to trace for recall any lot that might prove to be bad). If the bacteriological examination should reveal an unsatisfactory condition, future consignments would be held at the port until a sequence of tests (on say three consecutive consignments) proved satisfactory. In addition, for products receiving only indicator tests, examination might be waived except for say a fifth of the shipments, chosen at random. These systems have been in use by various countries.

When a food product is produced and shipped under uniform and adequate controls, a comparatively small routine sample (e.g. $n = 5$) can be used. Until such assurance has been obtained through good past history and/or surveillance of some sort, the only way to obtain satisfactory protection (i.e., substantial discrimination between good lots and bad lots), is by taking investigative samples (larger ns). But, as noted earlier (Section E (e), page 17), to double the reliability may require four times as large a population sample. The cost of the additional testing should be balanced against the potential gain in discriminating power, and a practical decision made.

Other questions about the lot, which often concern an analyst when assessing the sampling needs of lots from a factory unknown in quality of process and control practice, include the following: (i) in the instance of a very large lot consisting of several carloads or many large containers filled with small packages, how best to obtain representative samples? (ii) to what extent might n need modification, depending on the size of the lot, accessibility of sample units (see Section D, page 16), and degree of homogeneity of distribution of the test organism? Finally, the analyst should consider the effect of population sample size n on the discriminating power of a sampling plan, a question discussed in Chapter 3, Section F.

(i) Food commonly is shipped in carloads or large cases containing smaller packages or units. Sample as widely as possible and at random among the carloads or cases, and then sample randomly a few of the packages among those chosen; that is, the sampling should be 'stratified' (See Sections B, page 10 and E, page 17). This is better than choosing a few carloads or cases, and many packages within each such chosen carload or case.

(ii) It may not be feasible to sample at random over the entire consignment. It may, in fact, only be possible to sample randomly from a portion of the consigment called the 'frame' (see Section D, page 16). If so, then results apply only to the frame, not the consignment, and it is for the regulatory agent to judge how far the results of the frame can be trusted to apply to the entire consignment. For example, accessible containers may be those nearest the door of a vehicle, nearest the hatch of a ship's hold, or on the periphery of a stack in a warehouse. The sample units chosen from such frames may sometimes represent the parts of a lot or consignment which have been exposed to greater hazard, either of contamination or of multiplication of existing organisms, in which case the use of the 'frame' samples would provide greater consumer protection through the selection of those most likely to be hazardous. One thing working in favour of reliable results is that there often is unconscious or unplanned randomization in the whole process of production, packing, shipment, etc.

L WHAT IS A SATISFACTORY 'PROBABILITY OF ACCEPTANCE?'

The stringency of a plan is measured by the probability of accepting lots in which a particular proportion of sample units is defective. A relatively lenient 3-class plan suggested in this book ($n = 5, c = 3$), accepts a lot with a 5% proportion of defective units and 30% marginal units on about three occasions in four ($P_a = 0.75$). The most stringent 2-class plan ($n = 60, c = 0$) would accept lots with the same proportion (5%) defective on about one occasion out of 20 ($P_a = 0.05$); or lots with $^1/_2$% defectives on about 19 occasions in 20 ($P_a = 0.95$).

It might at first be supposed that the former plan provides no worthwhile protection at all; and that even the latter large-scale examination provides very little. This supposition is negated as soon as one considers the implication of such figures in practice. A probability of acceptance on three occasions in four ($P_a = 0.75$) means that one lot in every four will be decisively rejected, a loss clearly serious enough to impel a manufacturer to reduce the microbial content in his product to a level well below the limit(s) set in the test employed. Even the rejection of one lot in twenty ($P_a = 0.95$) should suffice to have this effect.

Nevertheless, the immediate protection conferred on the consumer as regards a particular lot is seriously limited when using practicable numbers of sample units; hence the recommendations to use large values for n when a direct hazard is recognized (as, for example, in Chapter 5, Section B on *Salmonella*).

The sampling plan, however, has also to be considered in the light of the associated criteria. Consider the example of a moderate health hazard, such as the presence of *Staphylococcus*. If the microbiological criterion m were placed at a numerical level well below that likely to represent hazard, one could frequently accept lots containing a high proportion of marginally acceptable units, which would require only a lenient sampling plan. If, on the other hand, m were placed nearer the

TABLE 11

Suggested sampling plans for combinations of degrees of health hazard and conditions of use (i.e., the 15 'Cases')

Degree of concern relative to utility and health hazard	Conditions in which food is expected to be handled and consumed after sampling, in the usual course of events[a]		
	Conditions reduce degree of concern	Conditions cause no change in concern	Conditions may increase concern
No direct health hazard Utility, e.g. shelf-life and spoilage	Increase shelf-life Case 1 3-class $n = 5, c = 3$	No change Case 2 3-class $n = 5, c = 2$	Reduce shelf-life Case 3 3-class $n = 5, c = 1$
Health hazard Low, indirect (indicator)	Reduce hazard Case 4 3-class $n = 5, c = 3$	No change Case 5 3-class $n = 5, c = 2$	Increase hazard Case 6 3-class $n = 5, c = 1$
Moderate, direct, limited spread[b]	Case 7 3-class $n = 5, c = 2$	Case 8 3-class $n = 5, c = 1$	Case 9 3-class $n = 10, c = 1$
Moderate, direct, potentially extensive spread[b]	Case 10 2-class $n = 5, c = 0$	Case 11 2-class $n = 10, c = 0$	Case 12 2-class $n = 20, c = 0$
Severe, direct	Case 13 2-class $n = 15, c = 0$	Case 14 2-class $n = 30, c = 0$	Case 15 2-class $n = 60, c = 0$

a More stringent sampling plans would generally be used for sensitive foods destined for susceptible populations.
b See 'Conclusions,' p. 39, for explanation of extensive and limited spread.

dangerous level, one would wish to accept such lots only seldom, requiring the use of a relatively stringent sampling plan. If *m* were at a dangerous level, one would not wish to accept lots containing any units exceeding that level, and 2-class plans of high stringency would be required. Adjustment is apparently possible by choosing limits known by experience to be associated with safety. When this is done, even though a high proportion of lots with substandard units will be accepted, the probability of consuming food that would cause illness is kept low.

Where more than one test is applied, the stringency of the examination is increased, to a degree depending on whether each test is independent of the others. With completely independent tests, the number of lots rejected would be the sum of the number of lots failing to meet the separate acceptance criteria for each. With completely interdependent tests, no change in acceptance would result, because failure to meet either test would be cause for rejection. This suggests, for example, that where there is an adequate direct test for a pathogen, there is no additional advantage in testing also for other 'indicator' organisms assumed to indicate the presence of that pathogen. In practice there is far from complete interdependence between different tests, and hence, in general, multiple testing imparts greater stringency.

Where seriously dangerous organisms are in question, so that even an occasional cell represents an intolerable risk (e.g., *C. botulinum* in some canned foods), no conceivably practicable system of examination could afford sufficient immediate protection to the consumer. In such cases safety must be sought in procedures other than direct bacteriological examination (see Chapter 6).

M SELECTING *n* AND *c*

The choice of *n* and *c* varies with the desired stringency (probability of acceptance), and thus with the cases in the grid of Table 6 (page 33). For stringent cases, *n* is high and *c* is low; for lenient cases, *n* is low and *c* is high. Table 11 shows a series of suggested sampling plans, which demonstrates this principle and is meant as a general guide to selecting plans. Each situation may, of course, be considered without reference to Table 11. The choice of *n* is usually a compromise between what is an ideal probability of assurance of consumer safety, and the workload the laboratory can handle. Consider first the nature of the hazard, then decide the acceptable probabilities of acceptance for the hazard in question.

Figure 3 (page 54) has illustrated the decision steps needed to choose between 2- and 3-class plans. For the presence or absence of a pathogen we necessarily use 2-class plans. Frequently the most desirable situation in which food would be free of a pathogen (implying a 2-class plan with $c = 0$) is not attainable even under GMP. Control agencies must, therefore, use judgment in applying a severe restriction of this kind. For example, present technology does not always produce *Salmonella*-free raw meats and poultry. If the agency, after judging the risk/benefit, is willing to accept positive units, a 3-class plan may apply. If, on the other hand, it is not willing to accept any positive units, a 2-class plan must be used. All 3-class plans proposed in this text have c values greater than zero since it is believed that no hazard, either of disease or spoilage, will occur when the foods in question have the specified numbers of results between m and M. Indeed, if $c = 0$, then $M = m$ and we have a 2-class plan.

Once it has been decided whether 2-class or 3-class plans will be used, and what OC curve is desired, the appropriate table is consulted to determine the values of n and c that will give the desired protection. For a 2-class plan, with $c = 0$, choose n from Table 2 (page 22); for a 2-class plan with $c = 1$ or more, choose n and c from Table 3 (page 23); for a 3-plan, use Table 4 (page 26).

If the number of sample units n so chosen exceeds laboratory capacity, reduce the types of tests applied, or reduce n, which will mean increased probability of acceptance of bad lots for given c.

By adjusting the criteria of acceptance in the test (decreasing c), or by increasing the sensitivity of the sampling plan by increasing the size of the total sample (either the number, n, or the size of the sample units analysed), the stringency of the examination as a whole can be increased. In this way, pressure can be brought to bear upon the standards of sanitation, nature of purchasing specifications, severity of processing and extent and nature of the quality control, practiced within the food industry concerned. The desirable effect should be carefully weighed, and the decision made should be known and understood by producer, manufacturer, and control agency alike. This has seldom been the case hitherto.

In summary, the final judgment on which a sampling plan is to be based should involve the relative weight to be placed on the foregoing microbiological, epidemiological, and ecological factors as well as the statistical probabilities of acceptance or rejection desired, and the economic considerations arising within the laboratory (e.g., limited physical facilities, equipment, and personnel). Some degree of subjective judgment is unavoidable at present, because data adequate to allow fully objective decisions take a long time to accumulate. Under such circumstances, individual judgments can vary widely, and if applied to international plans could give rise to serious confusion. For this reason the Commission has spent much time in collaboration with international consultants to reach the collective agreements expressed in the plans recommended in Part II.

N ROUTINE AS OPPOSED TO INVESTIGATIONAL SAMPLING

While the chief concern of this book is the use of sampling plans for routine examination, circumstances of unusual hazard require more intensive, 'investigational,' examination. Most of the plans recommended in Part II are designed for routine examinations. Presumably they would fulfil their safeguarding rôle whether or not the food was from a known manufacturer, on the assumption that the routine examinations would detect severe fault. Such a result might itself be a cause for more intensive 'investigational' study if the food were from an unknown source. For example, if abusive treatment were detected by the routine plan, additional capacity to detect pathogens would be required, particularly if food from the offending source were already distributed within the importing country. Alternatively, where the evidence of hazard was strong, the test agency might refuse entry without even analysing the food, or might require reprocessing.

Specific investigational plans are not proposed, because the circumstances warranting the use of more stringent plans are so variable, and because most food-testing microbiological laboratories are already working near capacity. The scale of any increased analytical effort will depend critically upon the competence and other priorities of the individual laboratory, and on the severity of the anticipated hazard. Under conditions of severe risk, for example if the concern were a critical pathogen (e.g., *S. typhi*, *C. botulinum*, *V. comma*) or if disease of unknown origin were encountered, it would be normal for the test agency to undertake studies to the limit of its resources, or to organize co-operative studies among several agencies or departments.

Table 12 compares the circumstances which, respectively, warrant the use of routine or investigational examination. The choice may require some measure of subjective judgment (based on experience), depending on the number and kind of factors which point to undue hazard. The evidence should be appraised *in toto* before deciding on the appropriate intensity of examination.

The investigational examination embodies two principles: (i) the use of more stringent sampling plans, thereby increasing the probability of detection of the test organism(s), and hence reducing the consumer's risk; (ii) extending the tests to include direct examination, wherever appropriate methods exist, for all relevant pathogens or toxins. In

TABLE 12

Choice of plans for routine or investigational sampling of foods

Circumstances related to the test food	
Warranting routine sampling	Warranting investigational sampling[a]
A THE FOOD	
Previous test satisfactory	Prior tests frequently unsatisfactory
Indicator tests give no evidence of severe contamination	Routine tests have revealed severe indicator contamination
Not commonly involved in disease	Of a type frequently involved in disease outbreaks
Not specifically suspect in an outbreak	Food from same manufacturer currently involved in a disease outbreak
	Of a type known to provide serotypes newly causing disease
	Circumstances point to possible involvement in an outbreak
Not primarily destined to sensitive populations	Suspect and destined to a sensitive population
	New type of food or new formulation with some rationale for hazard
	Examination results from different laboratories in conflict
B THE MANUFACTURER	
Records satisfactory	No records
Sanitation control known to be normally adequate	Known or suspected to exercise unsatisfactory plant-control
	Importer has knowledge of temporary hazardous circumstances at the plant
C COUNTRY OF ORIGIN	
Known to exercise competent control of plant practice	Endemic or epidemic situations hazardous
Not in areas endemic or currently epidemic for pertinent food-borne disease	Carrier rate high
	Sewage pollution usually severe
	Food control systems primitive

a To be appraised *in toto*. Not all individual factors would, alone, necessarily warrant investigational sampling.

practice this means increasing, as much as possible, the size of the individual sample units and also the number of sample units n, although a relatively large increase in n is necessary to effect substantial increase in the sensitivity of a sampling plan (as noted on page 21). Thus, consumer safety is best protected by direct examination for the pathogen(s) of concern, with samples as large as feasible (see this Chapter, Sections H and M). This can be aided by aggregation of sample units (page 70).

5

Sampling plans for situations involving direct hazard from pathogens

Judgments about hazard from pathogens are based largely upon bacteriological analysis, but the results of such analysis, and conclusions derived therefrom, depend on the sensitivity of the method used.

Early methods of detecting *Salmonella* in faeces or blood of a diseased person were not really suitable for food analysis. When more sensitive or discriminating media were used, some foods previously thought to be generally free of *Salmonella* frequently proved to be contaminated. Highly sensitive methods have not yet been developed for *V. comma*, *Sh. dysenteriae*, several zoonotic bacterial pathogens, or viruses, in foods. Hence, opinions on frequency of contamination by such species may need review as more sensitive procedures become available, or if more 'durable' strains arise, e.g., the El-Tor strain of *V. comma*.

The sensitivity of examination for *Salmonella* has been improved further by increasing the quantity of material actually examined. Previously, it was customary to examine samples of food weighing 1 g, or 10g; now it is usual to examine 25 g, or 50 g, occasionally even more. The importance of this will become evident below.

A THE IDEA OF ZERO TOLERANCE

In the past, it has been customary to specify that pathogens 'shall be absent' from foods, and the Commission proposed that for them no numerical tolerance should be expressed. At least three considerations challenge this ideal:

(i) No feasible sampling plan can ensure complete absence of a particular organism (Chapter 4, Section L). Even when $c = 0$ (i.e., the lot will be rejected if the pathogen is found in any sample examined), it

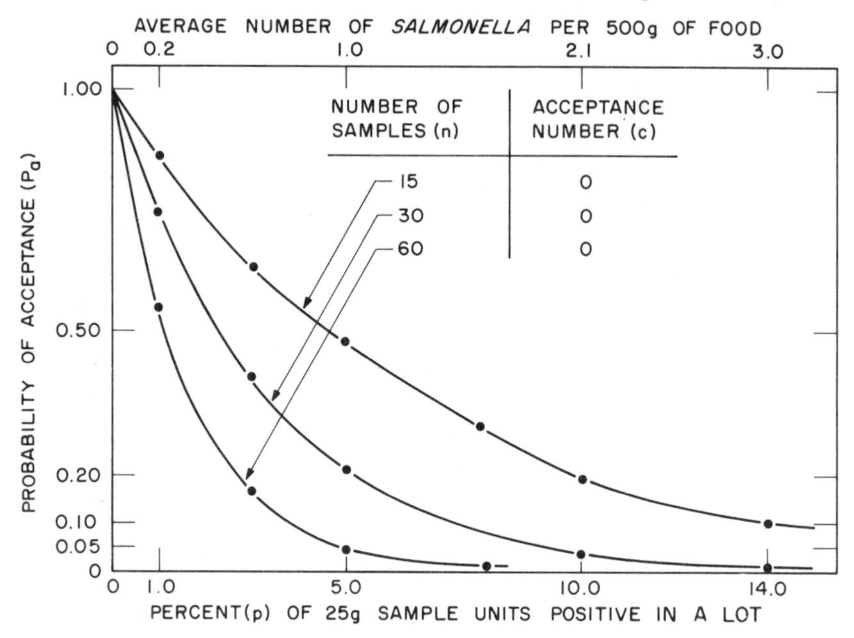

Figure 5 Relations between the sampling plan and the degree of security conferred – OC curves for three sampling plans proposed for *salmonella*

cannot be guaranteed that the lot is completely free of the organism, no matter how large the number of sample units n. However, what can be found is the probability of acceptance (P_a) for lots of various qualities, as a function of the given n and c. (For example, $n = 60, c = 0$ has $P_a = 0.5$, i.e., one chance in two of being accepted, for lots in which 1% of the sample units are defective, i.e. $p = 0.01$, see Fig. 5, above).

(ii) Plans in which $c = 0$ are not necessarily the most exacting. For example, if one sets a limit of 5% defectives in a lot, the plan $n = 95, c = 1$ will accept 'bad' lots ($p > 0.05$) less often, and 'good' lots ($p < 0.05$) more often, than will $n = 60, c = 0$. In other words, the former plan is the more discriminating, even though it admits the presence of the pathogen. Preference for $c = 0$ is infiuenced by the wish to emphasize that absence is the desired objective (although it cannot be guaranteed), and by the knowledge that pathogens such as *Salmonella* may be found by direct examination of certain foods, and having been demonstrated, are difficult to overlook.

(iii) It is not yet commercially possible to market some foods completely without pathogens. For example, most raw meats contain *Salmonella* at relatively high frequency, and no immediate practical solution is in sight. A sampling plan adjusted to this situation would seem more realistic and satisfactory than one based on an unmeasurable ideal of complete absence (i.e., zero tolerance).

The need for compromise

The foregoing illustrates the difficulty of trying to make a rational compromise between a desire to eliminate pathogens completely to protect consumers, and what have been considered practicable methods of production. In this field, past decisions have been arbitrary, and some of them illogical. Thus, no restriction is currently enforced on the enormous quantities of poultry and carcass meats which often are regularly contaminated with *Salmonella*, though these foods are clearly shown by epidemiological evidence to be major causes (direct or indirect) of *Salmonella* food-poisoning. On the other hand, drastic and costly action has been taken on finding *Salmonella* in sugar confectionery and chocolate, though few outbreaks due to such products have been reported. The risk entailed by the regularly contaminated foods is probably as great as that entailed by an operator who is an infective carrier and whose detection would lead to his removal. Therefore, sampling procedures are needed which remove the more highly contaminated lots without completely strangling production.

To encourage a realistic choice of sampling plans for hazardous pathogens, and because of the current international effort to control *Salmonella* in foods, the problem is considered in further statistical detail with particular reference to *Salmonella*. Similar considerations might conceivably be applied, for example, to the *Arizona* group, enteropathogenic *E. coli*, or *Shigella*.

B SAMPLING PLANS FOR *Salmonella*

Any member of the genus *Salmonella* presents some degree of hazard to human health. Some, like *S. gallinarum* and *pullorum*, are of minor significance for humans; other, notably *S. typhi*, are a severe hazard. But the majority, to which this discussion relates, represent a moderate direct hazard which, through cross-contamination and multiplication, can be spread readily via foods.

The numbers and incidence of these organisms can and should be low. Foods entailing a minimal disease hazard are examined for *Sal-*

monella only when there is unusual cause for concern. Such examination is needed to decide whether a product is acceptable in any of the following circumstances: (a) the food is of a kind commonly involved in salmonellosis; (b) the food is intended for a sensitive population (see page 41; and Table 8, page 42); (c) there is no knowledge of the manufacturing processes, or of the bacteriological record of earlier consignments; (d) there is special reason to suspect that the product of a particular factory may contain *Salmonella*.

Because most food control efforts should be directed to the areas of greatest risk, foods are assigned to categories according to the degree of hazard they present (see Tables 6, pages 33 and 11, page 60). To each category is attached an appropriately stringent method of deciding the acceptability of the finished product. Case 10 recognizes a relatively mild hazard such as exists when a food may contain *Salmonella* but will be used by ordinary consumers after cooking (e.g., red meats, poultry). Case 12 expresses a greater risk, where the conditions of use may favour spread and/or multiplication, and hence needs more stringent sampling. Case 11 is intermediate. Because absence of *Salmonella* is the preferred criterion of acceptance ($c = 0$), the stringency of examination is normally increased by raising the number of sample units n (see page 62). Table 13 shows recommended sampling plans for the various categories, with degrees of stringency related to hazard expressed by cases.

Our recommendations in Table 13 are similar in principle to those of the report of the *Salmonella* Committee of the US NAS-NRC (1968), in that both increase the stringency of the sampling plan with increase in the degree of hazard caused by the conditions of food use. An obvious difference is that the present proposals involve less stringent plans. For this there are two reasons:

(i) The sampling plans proposed by NAS-NRC were not intended for routine examinations, but for use when the question arises whether a product contains *Salmonella*. Routine examinations will from time to time raise this question, and the NAS-NRC plans were intended for the special investigation which it entails. For such investigations, the Commission agrees (Chapter 4, Section N) that plans are desirable in which the degree of stringency is as great as the NAS-NRC recommended (see Table 13).

(ii) If a few highly pathogenic types such as *S. typhi* are regarded separately, the remaining *Salmonella* types which occur commonly in foods (to which Table 13 refers) usually involve mild to moderate disease risks. The NAS-NRC report does not make a similar distinction and considers all types of *Salmonella* together, implying a high degree of

TABLE 13

Stringency of sampling for *Salmonella* for foods with different categories of disease hazard (all 2-class plans with $c = 0$)

ICMSF				NAS/NRC[a]	
		n values[b]			
Hazard[c] in use	Case[c]	Normal[c] routine	Special[d]	Category	*n* values[b]
Reduced	10	5	15	B	13
Unchanged	11	10	30	C	29
Increased	12	20	60	D	60

a Based on the Report of the *Salmonella* Committee US NAS/NRC (1968)
b n = number of sample units
c Compare Tables 6 (p. 33) and 11 (p. 60)
d E.g., where the food will be eaten by susceptible consumers (Table 8, p. 42), or for investigational purposes (Table 12, p. 62).

risk. The Commission agrees that where circumstances (e.g., unusual susceptibility in the probable consumers) create a higher risk, sampling plans of higher stringency are desirable (compare page 41; and Table 8, page 42).

Figure 5 gives the operating characteristic curves which express the relation between the degree of security conferred and the sampling plans proposed. For instance, consider the curve for the plan $n = 30, c = 0$. If the proportion of sample units defective (p) is 0.05 (i.e., 5% are defective), the probability of acceptance (P_a) is about 0.25, i.e., 1 in 4, hence 3 out of 4 such lots would be rejected. Suppose the units are 25 g samples – a realistic supposition with a food like dried or frozen egg. They are defective if examination shows them to contain any *Salmonella*. Hence, if there are 5% of defective units (i.e., one defective in 20) there is a *Salmonella* in $20 \times 25\,g = 0.5\,kg$ of food (assuming each contaminated unit to contain one cell). That is to say, the plan $n = 30, c = 0$, with unit samples of 25 g, will reject 3 out of 4 lots in which the level of contamination is about two *Salmonella* cells per kg.

It is only approximately correct to assume that each defective unit contains one cell: a few are likely to contain more than one. The difference is trivial for present purposes, however. For example, the plan $n = 60, c = 0$ has $P_a = 0.05$ (i.e., one in 20) for lots in which 4.87% of sample units are defective ($p = 0.0487$). When account is taken of the possibility of more than one *Salmonella* cell occurring in a sample unit, this corresponds to an average not of 0.0487 but of 0.050 cells per sample unit. If the sample units are of 25 g, this then corresponds to 0.050 cells per 25 g, or one cell per 500 g, on average. Thus to use $n = 60, c = 0$ with 25 g samples would reject 19 out of 20 lots containing two *Salmonella* cells per kg. If the sample units were 50 g each instead of 25 g, the corresponding cell concentrations detectable would be half as great, i.e. one cell per kg (see Section E(e), page 17).

To reduce laboratory effort while maintaining the stringency of the sampling plan, it would be helpful to combine either groups of sample units or the enrichment cultures therefrom, using plans in which the presence of a single positive rejects the consignment. A laboratory investigation undertaken for ICMSF (Silliker and Gabis, 1973) has indicated that different sized subsamples drawn from a single well-mixed laboratory sample of dried food give very similar results. A subsequent investigation (Gabis and Silliker, 1974) using foods of high moisture content (including eggs, poultry meat, meat, and meat products) shows comparable reliability in the use of composite samples; for example, examination of three composites of 20 × 25 g sample units has given the same result as examination of all 60 sample units individually (the total amount examined being 1500 g in both instances). This suggests that considerable reductions may be made in the cost of laboratory analyses, by compositing sample units. Alternatively, such compositing offers the prospect of large increases in the number of sample units, n, with corresponding increase in the stringency of examination, without correspondingly increasing laboratory effort. No reduction in the cost of collecting the necessary number of sample units can be expected, since sample units to be composited would still have to be chosen randomly.

Examination of the same total weight of material as above recommended, drawn as a smaller number of sample units of proportionately greater size, is expected to give a result of lower stringency because of less ability to reach random concentrations of organisms. Of course, stringency is diminished if the sample units are smaller than the 25 g discussed above, though the same in number (see Section E(e), page 17).

C PROBLEMS IN THE IMPLEMENTATION OF STRINGENT SAMPLING PLANS

(a) *Conceivable alternative plans*

As indicated above, for a pathogen like *Salmonella* there are strong reasons for adopting sampling plans in which $c = 0$:

(i) It is objectionable to some people to admit the presence of a recognized pathogen.

(ii) There is the very great technical advantage of combining many samples for bacteriological examination in systems with $c = 0$, because a single positive decides the outcome.

(iii) The plan with $c = 1$ (or more) would always require the examination of (many) more sample units, for equal probability of acceptance, than a plan with $c = 0$.

(iv) Even if the indicated probabilities are the same, their implications are not. For example, when a lot is accepted with the plan $n = 60, c = 0$, it is possible (though not certain!) that the lot may perhaps not be contaminated at all. But using the plan $n = 95, c = 1$ and accepting one positive, it is known that the lot certainly is contaminated (though the probability that a lot with 5% of sample units contaminated would be accepted is in both instances $P_a = 0.05$).

(b) *Fallacious procedures*

It might be suggested for example that, if the plans $n = 60, c = 0$ and $n = 95, c = 1$ provide equivalent probability (for lots having 5% of units defective), an operator finding one positive in 60 sample units might then proceed to examine another 35 (total $n = 95$) in the hope of clearing the lot if all the latter proved negative. But such a procedure is in reality a two-stage plan $n_1 = 60, c_1 = 1$, plus $n_2 = 35, c_2 = 0$, which has a greater probability of accepting an unsatisfactory lot than $n = 95, c = 1$. In fact, this probability P_a is 0.07, compared with 0.05 for the one-stage plan. Although in this example the difference is not great, there are situations where such two-stage procedures cause more serious error. Where two-stage sampling plans are actually being used, their OC curves should be computed and the resulting probabilities of acceptance evaluated.

A similar fallacy can arise where a plan requires a large number of sample units, which would be unusually costly. Suppose the plan is $n = 95, c = 1$, but for economy a group of only 20 units is tested initially. If one unit should fail to comply in this initial group, an analyst might examine the remainder (in this case 75) with the idea that if he does not find any defectives in the second group, he may ignore the first and concentrate on the finding of no defective out of 75 units. Nevertheless, the plan actually corresponding with the tests made is $n_1 = 20, c_1 = 1$ and $n_2 = 75, c_2 = 0$. No justification exists for 'preferring' the results from the second series. Similarly, if the plan were $n = 60, c = 0$, and the analyst felt uncertain about accepting his initial finding of a defective

TABLE 14

Comparison of some operating characteristics of several sampling plans relevant to examination for pathogens (data collected from Tables 2 and 3, pages 22–3)

	A For lots with $p = 0.10$, i.e., 10% of samples contaminated			B For lots with $p = 0.02$, i.e., 2% of samples contaminated		
Sampling plan n, c	P_a[a]	P_r[b]	Approximate proportion of lots rejected	P_a[a]	P_r[b]	Approximate proportion of lots rejected
$n = 60, c = 0$	<0.005	0.995	199/200	0.30	0.70	2/3
$n = 10, c = 0$	0.35	0.65	2/3	0.82	0.16	1/6
$n = 5, c = 0$	0.59	0.41	1/3	0.90	0.10	1/10
$n = 3, c = 0$	0.73	0.27	1/4	0.94	0.06	1/16
$n = 5, c = 1$	0.92	0.08	1/12	1.00	<0.005	$<1/200$

a P_a means probability of acceptance of lot
b P_r means probability of rejection of lot

unit in the first 20 sample units examined, and if the product were accepted on the basis of the second set of results, the actual plan applied would be the double sampling plan: $n_1 = 20$, $c_1 = 1$; $n_2 = 40$, $c_2 = 0$, which actually accepts more defective lots than would $n = 60$, $c = 0$.

D RELATION TO CURRENT COMMERCIAL PRACTICE

Stringent sampling plans like those just suggested above represent an ideal not yet attainable for products like poultry or carcass meats, which are found to be frequently contaminated by *Salmonella* when specially sensitive methods of detection are used. At present it is often recommended to use procedures like whole carcass rinsing or swabbing, where in effect the sample unit is most of the carcass, and a positive may correspond to levels of contamination of the order one *Salmonella* cell per kg for poultry, or even one per 100 kg for carcass meats. Using such criteria, one finds slaughter lines producing carcasses found to be contaminated with a frequency of 10% or more. For lots with 10% defectives (i.e., $p = 0.10$), the relations shown in Section A of Table 14 are valid. The sampling plan $n = 60$, $c = 0$ would reject virtually all such material; $n = 5$, $c = 0$ would reject one-third of it (Fig. 6). Even the plan $n = 5$, $c = 1$ would reject one in twelve lots, a commercially severe rate of rejection. If, as a result of rejections on the basis of $n = 5$, $c = 1$, the quality of lots improved to the point where only 2% instead of 10% of carcasses were found contaminated, the relations would be as in Section B of Table 14. If 2% of carcasses are contaminated, the indicated average concentration of *Salmonella* cells is of the order one per 50 kg for poultry, and one per 5000 kg or more for beef. If a 2% frequency of contamination were still thought too great, a similar pressure for further improvement (rejecting about one in 10) could be reintroduced by changing the sampling plan from $n = 5$, $c = 1$ to $n = 5$, $c = 0$.

This illustrates how, by periodical adjustment of the stringency of the sampling plan, a steady pressure to improve could be applied without placing catastrophic restrictions upon the supply of the commodity at any time.

The sampling plan $n = 5$, $c = 1$ has P_a = about 0.65 for lots with $p = 0.24$ (see Fig. 6), i.e., this plan would reject one out of 3 lots in which 1/4 of the units were contaminated. If these units were poultry carcasses weighing about one kg, 1/4 contaminated corresponds to a level of contamination of the order of one *Salmonella* cell per 4 kg of food; while with beef carcasses, the figure would be of the order of one per 400 kg. This kind of situation is very different from the examination of

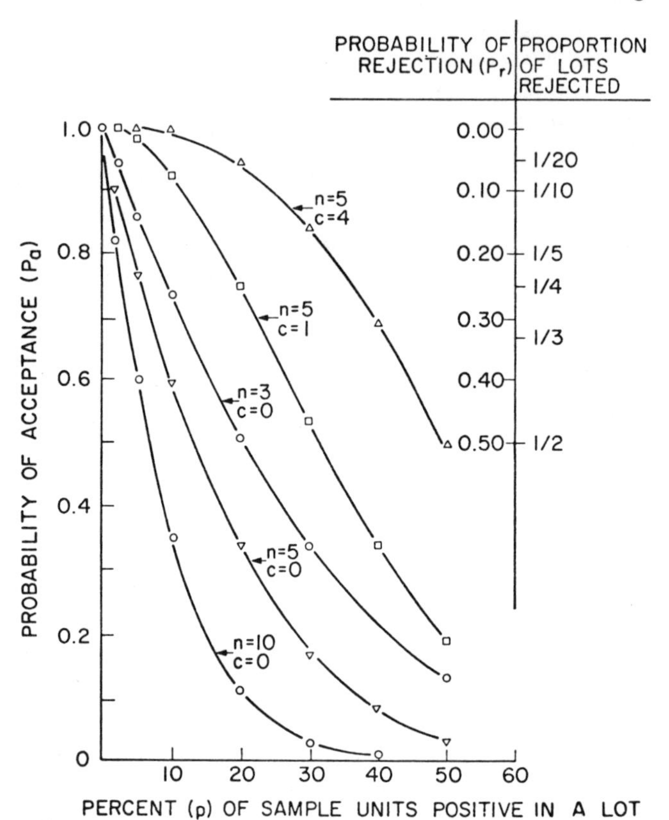

Figure 6 Operating characteristic curves for sampling plans with small numbers of sample units (data collected from Tables 2 and 3, pages 22–3)

sample units of order 25 g, e.g., of egg. The plan $n = 60$, $c = 0$ rejects about one-third of lots in which 0.7% of sample units are defective, which (these being 25 g sample units) corresponds to a level of contamination about 1 cell per 3 kg. That is to say, the plan $n = 60$, $c = 0$, applied even to 25 g sample units (e.g., of egg), actually provides less protection than the plan $n = 5$, $c = 1$ applied to whole carcass samples of poultry of effective weight about one kg; even the latter plan would reject a substantial proportion of current production of poultry carcasses.

In such circumstances, the immediate application of the more stringent sampling plans to poultry, and still more to larger carcasses, seems at present unrealistic. The application of even so loose a plan as $n = 5$, $c = 1$ would immediately require a substantial improvement on current levels of contamination; and that improvement could then be continued by proceeding to plans of successively greater stringency. It is important to understand the difference between the sampling plans for some raw meats and those for other foods as regards *Salmonella*, and to note the effect of the size of the sample unit on stringency of examination, as mentioned at the beginning of this chapter.

6

Control at source

From what has been said already, it should now be clear that any practicable system of bacteriological examination with a food product in commerce can never provide complete certainty that a desired bacteriological state has been attained: it can indicate this only with a particular degree of probability. Moreover, in particular cases (with workable schemes of examination), this degree of probability is either barely adequate, as in trying to detect *salmonellae* (see Section B, page 67), or is even totally inadequate, as in trying to detect *C. botulinum* in canned foods (see Section L, page 59). In such situations it is always wise to use other safeguards besides microbiological examination; and where that examination cannot assure reasonable probability of protection, these other procedures become the principal or only safeguard.

The best such safeguard is a process (e.g., appropriate sanitation, or heating) known to reduce or eliminate the relevant microorganism(s), the process being monitored by a system of examination to verify that it has been applied effectively. For this reason, control agencies often put a high proportion of their effort into 'Control at Source', where the essential purpose is to educate or compel management to follow 'good manufacturing practice' (GMP), derived from the joint experience of progressive manufacturers and the control agency. This requires that control be exercised over: the inspection and maintenance of equipment, the practice of sanitation throughout the plant, systematic microbiological tests for effective accomplishment at critical points in the processing scheme, and treatment of the finished product. Generally accepted codes of manufacturing practice should be followed by the processor, and used by control agencies as guides to assess adequacy of sanitation. Many countries have national codes of sanitary practice (Elliott and Michener, 1961; review of earlier codes), but of particular interest internationally are the FAO/WHO Codex Alimentarius Food Hygiene Committee's codes on General Principles for Food Manufacture and on Production and Marketing for Specific Food Classes, e.g. Processed Fruits and Vegetables. These are available in most countries from sources of FAO and WHO publications.

Statistical quality control techniques have been developed for this kind of 'in-plant' control. These include the use of accumulated data and control charts to show, before shipping the product, whether or not the process is 'in control.' They are outside the scope of this book, but details may be found in Burr (1953), Duncan (1965), Hamaker (1960), and Schaafsma and Willemze (1957).

Such 'in-plant' control gives greater security to the consumer by improving the economic efficiency of the control agency, which can never hope to have sufficient resources to test every production lot of foods. Foods known to have been produced under GMP may be received with confidence; and lots may thus be subjected to less demanding criteria or to less frequent analyses, than would those from an unknown manufacturer or from one known to practice poor quality control. Availability of in-plant records is also of great value, to both manufacturer and control agency, to help solve problems when a substandard or hazardous condition is found in the product.

All food manufacturers should therefore practice in-plant control, and should maintain appropriate records of test results. To help small companies to do this, Thatcher (1963), recommended that industrial associations should either organize a central testing laboratory, or co-operate with or support private professional testing laboratories.

Some import inspection agencies accept certification or inspection programs of the domestic inspection service in the country of origin. In this situation the domestic inspection agency becomes, effectively, an arm of the importing agency, and control at source is effective at long distance, provided the domestic inspection agency is competent and honest.

The alternative to control at source is a substantial increase of analytical control by official laboratories. The control agency frequently knows little or nothing of the producer of foods in international commerce. In these circumstances, testing of the final product is the only way to judge performance.

PART II SPECIFIC PROPOSALS FOR SAMPLING AND SAMPLING PLANS

Introduction

In the following ten chapters general directions for collecting and handling sample units are outlined, methods of sampling various food products are described, and sampling plans for various food commodities are given. The food commodities have been grouped into the following nine classes, each discussed in a separate chapter: fish and fishery products, vegetables, dried foods, frozen foods, milk and milk products, raw meats, processed meats, shelf-stable canned foods, and fresh or frozen raw shellfish.

Emphasis has been placed on those foods that are considered to be most in need of control analysis, and, hence, of acceptance sampling. This comprehends foods traded in large quantities internationally, and which are sometimes known either to cause food-borne illness or to suffer economically important spoilage. The order of presentation of the food classes in Chapters 8 to 16 has been chosen so as to proceed to foods in which sampling plans are only of special or limited use; indeed, for the final one considered (shellfish) bacteriological sampling of the product does not seem the best method of control, and no sampling plan is in fact suggested. This has been done to illustrate some of the limitations, as well as the advantages, of this type of procedure.

For every food listed, an expert committee has gone through a series of considerations to determine the appropriate 'case' (Tables 6, 9, 10, 11), microbiological criteria, and sampling plan.

Information on bacteriological content of foods and frequency of involvement in disease was sought from many sources. For some foods, too few data were available to establish microbiological criteria or a suitable sampling plan; and for all, the data were less than ideal. To establish appropriate criteria necessitates examination of a large number of analyses, which meant for our purposes the pooling of data from many different observers. It was seldom possible to take account of differences in sensitivity inherent in the use of different methods (see page 55). The microbiological criteria and sampling plans given here therefore represent only consensus judgments based on the total knowledge and experience of the Commission members and consultants. They are offered in the hope that they might be helpful on a worldwide basis, but they should be adopted as definite standards only with caution. The recommendations are subject to change as new data or experience may dictate. The Commission has already begun a project to compile, and examine by computer, data on food microbiology from around the world.

For some food commodities the sampling plans recommended are not consistent with the case in Table 11 (page 60), for one or both of the following reasons: (i) knowledge of the prevailing microbial content clearly showed that application of the plan appropriate to the specific case would lead to rejection of an impracticably high proportion of valuable food (e.g. meats, fish); (ii) testing laboratories lacked the resources necessary to examine the numbers of sample units required for the more desirable plan. Such a compromise should be offset by use of investigational sampling where needed.

The report prepared by each subcommittee (see Appendix 3, page 168) was examined in plenary session, sometimes several times, to obtain a consensus.

The proposed plans are addressed mainly to official control agencies. Companies with their own facilities for bacteriological control, frequently apply tests similar or identical with those used by government agencies, before release of their products. However, just as official agencies may reduce the frequency of examination for companies with a known history of good quality, companies with knowledge of their own performance might well use different testing systems based on their own quality control data. These could involve less stringent sampling plans, or perhaps automatic control charts (see also CONCLUSIONS, page 158).

7

Collecting and handling field samples and sample units

A GENERAL CONSIDERATIONS

Correct sampling requires careful attention. The objective is to obtain a representative sample of the food and to submit the sample units to the laboratory in a condition bacteriologically unchanged from that existing at the time of sampling. Samples should be collected only by an authorized person properly trained in the appropriate techniques. He may be assisted by other persons responsible to him. All persons concerned should take appropriate measures to prevent, as far as possible, any contamination of either the food consignment or the sample units and any microbial growth or death within the samples during transport to the laboratory and during subsequent storage and handling.

The sampling directions described in this chapter are considered good practice and it is recommended that they be followed whenever possible. It is, of course, difficult to lay down fixed rules to be followed in every situation, and particular circumstances may render some modification desirable on certain points.

The directions given in this book are those which apply generally, in collecting, labelling, transporting, and storing samples, and in preparing them for laboratory analysis. Specific sampling directions relating to a given food or class of food are given in Chapters 8–16.

B FIELD SAMPLE *VS* SAMPLE UNIT

The field sample is the material collected; the sample unit is the material actually used in the analysis. The two may be the same; but preferably the field sample should consist of at least twice as much food as that actually required for analysis, to provide a reserve in case of accident, or later enquiry. In many situations, field samples will consist of unopened containers (e.g., hermetically sealed cans, consumer-size packages of frozen vegetables) holding many times the amount of food needed for the sample unit. The subsample taken from the field sample in the laboratory is referred to throughout this book as the sample unit. The probability data given in Tables 2 to 5 relate to this portion because it is the amount of food actually used for the microbiological tests.

Each field sample provides only one result for each test carried out. Therefore, if more than one sample unit is tested from the same field sample, the results are averaged. The food homogenate prepared from one sample unit (Thatcher and Clark, 1968, page 60) will often provide enough material for a number of different microbial tests, but occasionally it will be necessary to take two or more sample units from the same field sample. For example, when tests for coliform and *Salmonella* are being run on the same food, one sample unit is required for the preparation of the homogenate (for the coliform test) and two others for *Salmonella* enrichment.

C GENERAL DIRECTIONS

(a) *Materials*

1 *Sample containers* Use clean, dry, sterile, leakproof containers such as wide-mouth glass or plastic jars or bottles, stainless metal cans, or disposable plastic bags, whose capacity is adequate for the sample unit desired – a minimum of 8 oz (227 g). Containers to be reused should be of a quality suitable for repeated sterilizations (see (b) below). Use jars, bottles, and cans with screw-type caps. Insoluble non-absorbent liners are necessary; impervious liners made of pressed pulp faced with vinylite are satisfactory. Moulded rubber, plastic, or cork stoppers may be used, but should be covered with an inert material such as aluminum foil or a plastic sheet before they are pressed into the sample container. Disposable plastic bags must be sealed securely with string or wire ties after filling so that they cannot leak during normal handling.
2 *Sampling devices* sterile triers, probes, drills, spoons, scoops, stirrers, pipettes, swabs, as required.
3 *Instruments for opening food packages* sterile scissors, knives, can openers, as required.
4 *Labels or markers* a light-coloured waterproof cardboard tag with reinforced eyelet hole and wire or cord ties; gum-backed paper labels;

felt-tipped indelible markers or adhesive-backed tape. Labels should be big enough to carry all relevant data (see (e) below).
5 *Sterilizing equipment* an autoclave, an oven capable of reaching a temperature of 170° c, gas sterilization chamber, and, for field sampling, either a 20–30 litre pressure cooker, a steam chamber, or a hydrocarbon (propane or butane) torch or burner, as necessary.
6 *Refrigerator* capable of chilling samples to 0–5° c in a few hours. A deep-freeze compartment is desirable for storage of frozen samples, which spoil when thawed, but a separate deep-freeze unit is necessary if many frozen samples are handled.
7 *Insulated container* a foamed plastic box or other insulated container suitable for transporting and/or holding frozen or chilled samples.
8 *Sterilizing agent* alcohol, 70% ethyl or isopropyl.
9 *Metal dial-type thermometer* −20° to 100° c with graduation intervals of not less than 2° c (and a spare). Optionally, use a shorter scale range if the range includes all temperatures to be measured.

(b) *Sterilizing*

1 Sterilize all sample containers and utensils that are to come in contact with the food. As a rule, sterilize beforehand in the laboratory by one of the following methods:

a autoclaving at 121° c for 15 minutes. Use for all materials containing water, and for dry materials likely to be damaged by dry heat (e.g., rubber);
b exposure to hot air in an oven at 170° c for at least one hour;
c ethylene oxide with carbon dioxide. Useful for heat sensitive plastics but requires special precaution in use.

2 Bring enough pre-sterilized utensils (triers, spoons, drills, etc.) from the laboratory to collect the desired number of samples; or clean and sterilize used ones at the site using one of the following methods:

a autoclave in a small (20–30 litre) pressure cooker at 121° c for 15 minutes;
b expose to steam at 100° c for one hour;
c heat in a small portable hot air oven at 170° c for one hour (assuming power is available);
d flame thoroughly with a propane or butane torch or Bunsen burner (taking care not to overheat sampling equipment);

e immerse in ethyl alcohol (70% v/v) and flame to burn off the alcohol (does not destroy bacterial spores);

f immerse for at least 30 seconds in a hypochlorite solution containing not less than 100 ppm of available chlorine or other acceptable halogen sanitizers which are bactericidally equivalent; rinse in sterile water and dry with sterile cloth prior to use.

Do not use (*d*) *or* (*e*) if there is danger of fire or explosion.

(c) *Collecting field samples*

1 Only an authorized person properly trained in the appropriate techniques should collect samples (see Section A, this chapter).

2 If possible, collect samples during unloading of a cargo of food (ship, truck, airplane) from all parts of the load ('frame'; see Section D, page 16). Be alert to the possibility of stratification. Take several times the amount actually required later for analysis in the laboratory (i.e. in most situations, about 200 grams) to provide a reserve portion in case of accident or later enquiry. Store the reserve under conditions in which its microbial condition will not change appreciably.

3 Submit samples to the laboratory in original unopened containers whenever possible. This will reveal the condition of the product as prepared and, in the case of finished commodities, as offered to the public.

4 If all individual packages are accessible, choose those to be sampled by use of random numbers (see Section C, page 11).

5 In sampling larger cartons of smaller packages, choose the desired number of cartons at random, and from each of these, again at random, the desired number of smaller packages. At each stage, use random numbers (see Section C, page 11). Preferably, take only one package from each carton; this procedure provides the most representative sample but may not always be feasible (see Section D, page 16).

6 If the product is in containers too large to be easily transported to the laboratory, transfer representative samples to separate sterile containers under aseptic conditions. Remove surface contamination on appropriate areas of the food package and the sample container by wiping with 70% alcohol, after removing gross dirt by washing or sponging with 70% alcohol or, in the case of paper containers, by removing the outer ply. Open sealed food packages carefully with a sterile cutting instrument such as scissors or a knife, as appropriate. Use a separately sterilized instrument for each package to avoid cross-contamination. Exercise special care where the risk of cross-contamination is high, e.g., with

light powders or pressurized packages. Do not open hermetically sealed cans under field conditions. Submit such cans to the laboratory as field samples (see item 3 above and Section B of this chapter).

7 If the food is in bulk, sample from various places within the container(s) rather than, say, just from the top or bottom (unless experience assures complete mixing, or unless sampling from some particular place(s) has become accepted as a result of long experience).

8 When sampling from an outlet or sampling port of a bulk container, allow some material to pass through to 'flush' the port before collecting the sample.

9 Avoid contamination in taking a sample from bulk or from a food package and when transferring the material to the sample container.

10 For collecting samples use the instrument appropriate to the physical state of the food; for example, triers for sampling blocks of hard cheese, drills for frozen products, scoops or probes for dried foods, pipettes for liquids, etc. (see recommendations for particular foods in Chapters 8 to 16).

11 Strive always to obtain a representative sample. Where possible, agitate liquid, suspended or free-flowing materials until the product is homogeneous. For unwieldy brittle materials such as frozen liquids, place material in a large sterile container, break it into smaller pieces under aseptic conditions (e.g. in a strong plastic bag), and take a sample of crushed material.

12 Despite the above instructions to sample aseptically, the inspector may sometimes send to the laboratory a sample obtained without asepsis. For example, boneless meat in bulk has many internal surfaces. The inspector can saw a large chunk from the frozen block, and send the chunk frozen to the laboratory. The microbiologist can take an aseptic sample of the internal surfaces for his sample unit. Consult the laboratory director before taking samples without asepsis, and inform the laboratory receiving the sample that the sampling was not aseptic.

13 Record the air temperature of the storage room or shipping vehicle as well as the temperature of the food from which the sample is drawn. Do not insert a thermometer into the food until after the sample has been taken. In instances where small packages are sent unopened to the laboratory, record the temperature of the food in an adjacent package in the same case or carton.

(d) *Number of sample units*

Draw the number of field samples equivalent to the number of sample units specified in the sampling plan used. See Chapters 8 to 16 for recommendations for individual foods and cases; also Section B, page 10 and Section C, page 11 for recommendations on random sampling, and Section B this Chapter for recommendations on the amount of food to take for each field sample.

(e) *Labelling sample containers*

1 Label all sample containers immediately before or after the sample is taken. Fix the label to prevent accidental removal during subsequent handling.
2 Number the sample label, and make a related record including all the information necessary to identify the sample (see (h) below).
3 When the field sample is taken from a larger container such as a bag or carton, identify this container with the corresponding number of the sample, in the event a subsequent examination is required.

(f) *Sealing of sample containers*

In critical situations, e.g., where the food is involved in a dispute or is suspect in a food-poisoning outbreak, it may be necessary to place an 'official' seal on the container of the sample unit, preferably in the presence of persons authorized to represent all parties concerned. The seal may be glued paper or other material which makes it impossible to tamper with the contents or the label without irreparably damaging the seal. Identify the seal with the date, sample number, and mark of the person collecting the sample. Where, for official purposes, multiple samples are required, treat each sample identically.

(g) *Transporting and storing samples*

1 All practicable means for handling a food change its microbiological condition, more or less. Therefore, examine samples immediately after they are taken, if possible. Otherwise choose conditions which minimize the possibility of change during transport and/or storage of samples.
2 Transport samples to the laboratory for analysis as rapidly as feasible. If the product is canned (shelf-stable) or is in a dry condition, take no particular precautions except to avoid temperatures above 45°C.
3 If the product is perishable and unfrozen, cool samples to 0–5°C and transport them in a protective insulated container (see (a), 7 above). Cool the samples in a refrigerator or more rapidly in an ice bath, and transport them in drained ice. Use waterproof sample containers if necessary. Preferably transport samples in such containers under direct

personal supervision to ensure that the containers will be held vertically and that liquid refrigerant in direct contact with sample containers will not rise to the level of the caps.

4 Maintain samples of frozen products in the frozen state until analysed. Collect such samples in prechilled containers and place in a freezer immediately after collection until their temperature is well below 0°C. Then transport them to the laboratory in an insulated container. Use dry ice (solid carbon dioxide) as the refrigerant if the time spent in transport may lead to thawing. If necessary, store in a freezer compartment or cabinet at about −10°C.

(h) *Sampling report*

1 Complete a detailed sampling report, signed by the person responsible for collecting the samples, and countersigned by representatives of the other parties concerned, if they are present. This report should provide the following information relevant to each lot in the consignment:

a name and address of the person collecting the samples;
b names and addresses of appropriate representatives of the parties concerned;
c date, place, and time of sampling;
d reason(s) for sampling;
e nature of the food;
f names of the manufacturer, importer, seller, buyer, as appropriate;
g number and size of the units constituting the lot;
h number and marking of the lot;
i type and identification of the transport vehicle;
j origin of shipment;
k place of destination;
l dates of shipment and arrival of the lot;
m number of the bill of lading or contract;
n method of sampling (random throughout lot, random throughout accessible units, or otherwise);
o size, number, and reference numbers of field samples;
p temperature of the product at the moment of sampling;
q means of transporting samples to the laboratory and by whom;
r name and address of laboratory analysing samples;
s tests to be made (e.g., SPC, *Salmonella*, *C. perfringens*).

2 The report should also include information on factors, conditions, or circumstances which may have influenced sampling or be pertinent to the investigation; for example, the condition of the individual packages, breakage, sanitation, humidity, etc.

(i) *Preparing sample units for analysis*

This involves withdrawing a subsample constituting the sample unit in a manner which will make it as representative as possible of the field sample from which it came (see Section B, page 10).

1 Begin the analysis as soon as possible after the field sample arrives at the laboratory.

2 If the field sample is frozen, use one of the following methods: (a) Partially thaw it in its original container (or in the container in which it was received at the laboratory) for 18 hr in a refrigerator at 2–5°C. (b) If the frozen sample can be easily comminuted, proceed without thawing. (c) With easily thawed material (e.g., frozen egg drillings), thaw in an incubator at 35°C for not more than 15 minutes.

3 Agitate thoroughly any food that is liquid, suspended, or free-flowing, before opening the sample container.

4 Avoid contamination when opening the sample container. Destroy surface contamination on and near the lid area of the container (or where the opening is to be made, in the case of foods in closed packages) by (a) flaming or (b) wiping with 70% alcohol and air drying or burning away the excess alcohol, as appropriate.

5 Take special care in opening sample containers or packages of powdered materials to avoid contaminating the working area and other samples. Avoid sudden movements and keep doors and windows closed.

6 When the product is composed of several components or layers (such as turkey and dressing, cream pie with meringue topping, etc.), it may be necessary to determine the microbial population or contamination of each part, for example, to reveal the possible source of an infection or intoxication. This is particularly important if the parts were processed separately, possibly giving rise to large numbers of microorganisms in some parts and not in others. In such cases, sample the parts separately while preventing cross-contamination as much as possible. In the case of cream pie, for example, take a sample of the topping without disturbing the filling and then, with a sterile knife or spoon, cut away the surface of the filling to expose filling uncontaminated by the surface layer or the meringue (Sharf, 1966).

7 Use aseptic technique in withdrawing and transferring samples to a tared blender jar.
8 For preparation and dilution of the food homogenate, follow directions given on pages 60–3 of Thatcher and Clark (1968).
9 Observe and record any abnormal appearance or aroma of the sample.
10 Store sample containers containing the reserve portion of the field samples in a refrigerator or a freezer as in (g) above, for replicate analyses if required.

8

Sampling plans for fish and fishery products

Fish and other sea foods include a very large number of different food items. The raw materials themselves are derived from animals from many genera or even phyla, thus giving rise to a diversity among sea food products which is not generally recognized. Such products are important articles in the international food trade. According to the recent FAO statistics (1969), over a quarter of the world's production of sea foods is exported. The order of importance, in terms of quantities exported, is shown in Table 15 along with figures for total production and the percentages exported. In 1969 the total amount of such products in international trade in the principal exporting countries amounted to 2,455,000 metric tons. The international nature of such trade may be judged from the fact that Canada, itself a fish exporting country, in 1970 imported fish and fishery products from more than 50 of the 150 producing countries.

A RELATIVE IMPORTANCE OF PATHOGENS, SPOILAGE ORGANISMS, AND TOXINS IN FISH

The potential for sea foods and other fishery products to act as carriers of food-borne disease is thus very large; yet all the data available, including national tabulations of food-poisoning statistics (Shewan, 1970; Bryan, 1971; Dolman, 1974), appear to show that such products, except molluscan shellfish, are relatively infrequent vehicles of infection. Only in Japan and probably in other areas of the Far East where sea foods constitute over 70% of the protein diet are they commonly incriminated in food-poisoning outbreaks (Sakazaki, 1969; Shewan, 1970; Thatcher, 1969). Sea water fish and crustacean shellfish are generally free from contamination with the common food-borne pathogens such as salmonellae and staphylococci at the time of harvest (Shewan, 1961;

TABLE 15

World total fish and shellfish and crustacean production, 1969[a]

	Class of product	Metric tons	% exported
I	Miscellaneous frozen fishery products (tuna, sardines, eels, trout, etc.)	4,636,000	12
II	Frozen fish fillets	648,000	60
III	Miscellaneous fish products (fresh or chilled)	364,000[b]	
IV	Fresh herring (chilled)	319,000[b]	
V	Miscellaneous fish products in air-tight containers	833,000	25
VI	Herring, sardines, in tins	506,000	37
VII	Molluscs (fresh frozen dried)	588,000	21
VIII	Crustacea (fresh frozen)	393,000	32
IX	Tuna, bonita in air-tight containers	394,000	19
X	Salmon in air-tight containers	93,000	53
XI	Fresh fish fillets	94,000	41
XII	Frozen herring	35,000	37
XIII	Fish products not in air-tight containers	1,530,000	1
XIV	Smoked herring (fresh frozen)	53,000	19
XV	Miscellaneous smoked fish products	367,000	1
	TOTAL	10,853,000	

a Excluding dried and salted fish but including smoked products (FAO, 1969)
b Quantity exported; no figures available for total production

Thatcher, 1969), unless fished from polluted water. But, like other foods, they can become contaminated during subsequent handling and processing and may acquire mesophilic spoilage organisms (growing at 35 to 37° C), *E. coli*, faecal streptococci, *Staphylococcus aureus*, and salmonellae. Newly caught fish may carry *Clostridium botulinum*, usually Type E, non-proteolytic Type B, and Type F (Craig and Pilcher, 1966; Craig *et al.*, 1968; Sakaguchi, 1969; Gangarosa *et al.*, 1971; Dolman, 1974), and also *Vibrio parahaemolyticus*. The latter species causes enteritis, particularly in the Far East where fish and fishery products are eaten raw or after various uncontrolled fermentation practices (Baross and Liston, 1970; Hechelmann *et al.*, 1971; Johnson *et al.*, 1971; Kampelmacher *et al.*, 1972; Sakazaki, 1969; Thomson and Thacker, 1972; Velimrovic, 1972). Both these organisms are usually present in low numbers and do not cause illness provided the foods are handled according to GCP. Their importance as agents of food-borne illness is confined to special cases.

Sampling plans selected for recommendation are for those products traded in large volume (Table 15). Formulation of microbiological limits was influenced by two major considerations: (a) the levels attainable in GCP, and (b) the possible food hazards involved.

Any realistic sampling plan should allow the inspecting agency to make the necessary examinations and to enforce the pertinent regulations without undue difficulty. This is particularly important with fish and fishery products where there is a large and continuing flow of yearly shipments, with a great diversity in the size, weight, and type of package. Usually the history of handling, processing, packaging, storage, and transportation is unknown. In addition, the perishability of many fish products requires that tests be completed quickly in order to minimize delay within the usual trade channels. With fish and fishery products, counts at 20 to 25° C are probably more valuable as indicators of incipient spoilage because of the psychrophilic nature of the spoilage flora. Hence we recommend that the Standard Plate Count (SPC) be carried out at 25° C. The SPC at 35° C is frequently of the order of one-tenth of the count at 25° C (Liston, 1957; Shewan, 1970).

Although in certain countries, GMP is commonly attained by plants processing sea foods, no information is available for a significant proportion of producers whose products are in international trade. The sampling procedures recommended take cognizance of this fact, the first consideration being the safety of the consumer. However, it is important that safety tests (e.g., for *Salmonella*) should not be confused with indications of shelf-life or quality, such as the SPC. Several items of fish and fishery products in international trade have SPC levels generally in excess of those proposed for some other foods (Table 16). There is no evidence, however, to suggest that they have given rise to any corresponding health hazard. Accordingly, the recommended SPC levels reflect presently acceptable practice in the fish industry, and are not related to criteria proposed for other dissimilar foods, with different records of safety. With cooked crustacea, however, while the SPC may be lower, contamination with coagulase-positive staphylococci has caused several food-poisoning episodes and, in this respect, a cooked fish product appears similar to other cooked foods with similar growth potential for pathogens.

In view of the increasing contamination of inshore marine waters with domestic sewage and hence of the increased probability of contamination of fish and crustaceans taken in such waters, examination for enteric viruses (including hepatitis) is probably at least as important as for *Salmonella* (Clark and Chang, 1959; Cliver, 1969; Kjellander, 1956; Woodward *et al.*, 1970). The current status of virus methodology prevents implementation of this (Cliver, 1969). Thus, methods for the isolation and identification of enteric viruses and of appropriate virus criteria for foods are an outstanding need in controlling food safety.

Other factors to be considered include ichthyosarcotoxin (tetraodon and ciguatera poisoning) and histamine poisoning. The former is due to naturally occurring toxins in the live fish (Halstead, 1965, 1967, 1970; Wills, 1967), whereas histamine poisoning is mainly due to postmortem bacterial action (Ferencik, 1970; Ienistea, 1971; Willis, 1967). It occurs almost exclusively in scombroid fish (e.g., tuna, mackerel, and some sardines) containing high levels of free histidine which is decarboxylated by bacterial action to histamine. The tests for both ichthyosarcotoxin and histamine are chemical in nature and are not included in this text but should obviously be considered in any decisions concerning sampling where fish are suspect in food intoxications.

Nearly all the data presented in Table 16 are concerned with marine fish and fishery products. In recent years there has been a considerable increase in the production of cultured fresh water species (FAO, 1970) and in the quantities and proportions of such products in international trade. There is every indication that this trend will continue and may in fact accelerate. In general, such products present a greater potential health hazard than do the other products given in Table 15. For this reason it is recommended that particular attention be paid to these products in the future.

B THE BASIS FOR SPECIFIC SAMPLING PROPOSALS

Only those fish products for which considerable microbiological data were available, derived from products from several countries, were considered for sampling recommendations. The available analytical data are tabulated in Table 16. These data were used in conjunction with analytical data from individual processing plants and from specific countries to assess attainment under current concepts of GCP. The values for m and M were assessed accordingly, based on knowing how bacterial content relates to acceptable quality and safety, and that these foods seldom cause food poisoning.

Except in the case of precooked products, the relation between SPC and safety or handling practices is not clear. SPC limits for unprocessed seafood products shown in Table 18 should be considered indicative and advisory only, and judgment should be made by a regulatory agency on a case-by-case basis on the application of these limits for rejection or seizure of particular products.

The stringency of the sampling plan was primarily dependent upon

TABLE 16

Existing data on the microbial content of fish products in international trade

Product		SPC at 25° C: Number per g				Coagulase positive staph per g				Faecal coliforms per 100 g				
		Sample units	< 10^5	10^5–10^6	> 10^6	Sample units	< 10^2	10^2–10^3	> 10^3	Sample units	< 100	100–360	360–3600	> 3600
Fish gutted and frozen at sea (cod)	Totals[a]	760	485	224	51	933	928	5	0	603	603	0	0	0
	%[b]		64	30	6		99	< 1	0		100	0	0	0
Fish fillets or frozen blocks	Totals[a]	15,993	8,290	6,908	795	3,568	3,545	18	5	11,939	10,008	1,742	47	142
	%[b]		52	43	5		99	1	< 1		84	15	< 1	1
Frozen comminuted fish	Totals[a]	595	187	166	242	90	87	3	0	688	452	52	58	126
	%[b]		31	28	41		96	4	0		67	7	8	18
Frozen freshwater fish	Totals[a]	3,757	1,445	1,709	603	250	247	3	0	3,421	3,033	240	142	6
	%[b]		38	46	16		99	1	0		89	7	4	< 1
Sliced smoked salmon	Totals[a]	49	32	14	3	52	52	0	0	48[c]	48	0	0	0
	%[b]		65	29	6		100	0	0		100	0	0	0
Cold-smoked marine fish (excluding kippers)	Totals[a]	1,093	858	141	93	917	917	0	0	352	328	11	4	9
	%[b]		78	13	9		100	0	0		93	3	1	3
Frozen kippers	Totals[a]	159	101	53	5	160	159	1	0	157	145	12	0	0
	%[b]		64	33	3		99	< 1	0		92	8	0	0
Frozen breaded fish portions	Totals[a]	2,721	1,314	1,223	184	1,371	1,364	7	0	2,144	1,986	102	49	8
	%[b]		48	45	7		99	< 1	0		93	5	2	< 1
Frozen fish sticks (fingers)	Totals[a]	1,863	1,245	548	34	1,007	1,000	7	0	1,817	1,516	133	130	38
	%[b]		67	31	2		100	< 1	0		84	7	7	2
Frozen fish cakes	Totals[a]	399	122	249	28	394	381	11	2	423	277	143	3	0
	%[b]		31	62	7		96	3	< 1		66	34	< 1	0

TABLE 16 (Concluded)

Product		SPC at 25° C: Number per g				Coagulase positive staph per g				Faecal coliforms per 100 g				
		Sample units	$<10^5$	10^5–10^6	$>10^6$	Sample units	$<10^2$	10^2–10^3	$>10^3$	Sample units	<100	100–360	360–3600	>3600
Raw frozen scallops	Totals[a]	980	665	262	53					978	888	49	22	19
	%[b]		68	27	5						91	5	2	2
Raw frozen lobster tails	Totals[a]	290	78	110	102	158	143	13	2	247	202	20	17	8
	%[b]		27	38	35		91	8	1		82	8	7	3
Raw breaded frozen shrimp	Totals[a]	483	37	159	287	230	221	7	2	1,486	1,060	392	34	0
	%[b]		7	33	60		96	3	1		71	27	2	0
Cooked frozen shrimp	Totals[a]	5,377	3,816	1,136	431	2,776	2,056	579	141	3,642	3,032	584	26	0
	%[b]		71	21	8		74	21	5		83	16	1	0
Cooked frozen lobster tails	Totals[a]	40	37	3	0					40	28	1	11	0
	%[b]		92	8	0						70	3	27	0
Cooked picked frozen crab	Totals[a]	2,242	1,624	441	177	4,632	2,839	1,730	63	2,528	2,066	333	129	0
	%[b]		72	20	8		62	37	1		82	13	5	0
Cooked frozen crustacea (lobster crab, shrimp)	Totals[a]	4,248	3,382	700	166	3,237	2,660	473	104	3,771	3,053	628	69	21
	%[b]		80	16	4		82	15	3		81	17	2	<1
Raw frozen shrimp	Totals[a]	2,990	714	345	1,931	1,467	1,395	58	14	3,024	2,499	248	247	30
	%[b]		24	11	65		95	4	1		83	8	8	1

a Total number of sample units tested and number of results falling within ranges indicated
b Percentage of total number of sample units falling within range indicated
c One sample was *Salmonella* positive

TABLE 17

Allocation of fish and shellfish products to case

Product	Condition	Test	Case
Fresh and frozen fish including fish frozen at sea, fish blocks, comminuted fish blocks, and scallops	(a) Cooked prior to consumption: hazard reduced	SPC	1
		Faecal coliforms	4
		Staphylococcus (ind.)[a]	4
	(b) Eaten without cooking, e.g., izuchi: health hazard increased	*V. parahaemolyticus*	12
Freshwater fish	(a) Normally cooked prior to consumption: hazard reduced	Utility	1
		Faecal coliforms	4
		Staphylococcus (ind.)[a]	4
	(b) Fish from warm waters	Faecal coliforms	4
		Salmonella	10
Cold smoked fish including kippered herring	(a) Normally cooked before consumption: health hazard moderate	SPC	1
		Faecal coliforms	4
		Staphylococcus (ind.)[a]	4
	(b) Eaten without cooking, e.g., salmon	SPC	1
		Faecal coliforms	6
		Staphylococcus (path.)[b]	9
Breaded pre-cooked fish products including fish sticks (fingers), fish portions, and fish cakes	Normally cooked prior to consumption, but heat treatment may be inadequate: low health hazard	SPC	2
		Faecal coliforms	5
		Staphylococcus (ind.)[a]	5
Frozen raw shrimp, prawns, and lobster tails	Normally cooked prior to consumption: hazard reduced	SPC	1
		Faecal coliforms	4
		Staphylococcus (ind.)[a]	4
		V. parahaemolyticus	10

TABLE 17 (Concluded)

Product	Condition	Test	Case
Frozen cooked shrimp, prawns, and lobster tails	Normally eaten immediately after thawing, but may be eaten after unpredictable delay (catered items): increased moderate hazard	SPC	3
		Faecal coliforms	6
		Staphylococcus (path.)[b]	9
		V. parahaemolyticus	12
Frozen raw breaded shrimp and prawns	Normally cooked or processed prior to consumption: hazard reduced	SPC	1
		Faecal coliforms	4
		Staphylococcus (ind.)[a]	4
		V. parahaemolyticus	10
Cooked picked crab meat	Not cooked prior to consumption; known health hazard: hazard increased	SPC	3
		Faecal coliforms	6
		Staphylococcus (path.)[b]	9
		V. parahaemolyticus	12

a (ind.) = considered as an indicator
b (path.) = considered as a toxinogenic pathogen

TABLE 18

Sampling plans and recommended microbiological limits for fish and fish products

Product	Test	Method reference[a]	Case	Plan class	n	c	Limit per g m	Limit per g M
1 (a) Fresh and frozen fish including fish frozen at sea, fish blocks, comminuted fish blocks, and scallops	SPC[b]	64	1	3	5	3	10^6	10^7
	Faecal coliforms (MPN)	77	4	3	5	3	4	400
	Staphylococcus (ind.)[c]	114	4	3	5	3	10^3	2×10^3
(b) Fish eaten raw	*V. parahaemolyticus*[d]	107	12	2	5[e]	0	10^2	—
2 (a) Freshwater fish	SPC	64	1	3	5	3	10^6	10^7
	Faecal coliforms (MPN)	77	4	3	5	3	4	400
	Staphylococcus (ind.)[c]	114	4	3	5	3	10^3	2×10^3
(b) Freshwater fish from warm waters	*Salmonella*[f]	90	10	2	5	0	0	—
	Faecal coliforms (MPN)	77	4	3	5	3	4	400
3 Cold-smoked fish including kippered herring (a) Cooked prior to eating	SPC	64	1	3	5	3	10^5	10^6
	Faecal coliforms (MPN)	77	4	3	5	3	4	400
	Staphylococcus (ind.)[c]	144	4	3	5	3	10^3	2×10^3
(b) Eaten uncooked	SPC	64	3	3	5	1	10^5	10^6
	Faecal coliforms (MPN)	77	6	3	5	1	4	400
	Staphylococcus (path.)[c]	114	9	3	5[e]	1	10^3	2×10^3
4 Breaded pre-cooked fish products including fish sticks (fingers), fish portions, and fish cakes	SPC	64	2	3	5	2	10^6	10^7
	Faecal coliforms (MPN)	77	5	3	5	2	4	400
	Staphylococcus (ind.)[c]	114	5	3	5	2	10^3	2×10^3

TABLE 18 (Concluded)

Product	Test	Method reference[a]	Case	Plan class	n	c	Limit per g m	Limit per g M
5 Frozen raw shrimp, prawns, and lobster tails	SPC[g]	64	1	3	5	3	10^6	10^7
	Faecal coliforms (MPN)	77	4	3	5	3	4	400
	Staphylococcus[c]	114	4	3	5	3	10^3	2×10^3
	V. parahaemolyticus[d]	107	10	2	5[e]	0	10^2	—
6 Frozen cooked shrimp, prawns, and lobster tails	SPC[g]	64	3	3	5	1	10^6	10^7
	Faecal coliforms (MPN)	77	6	3	5	1	4	400
	Staphylococcus (path.)[c]	114	9	3	5[e]	1	10^3	2×10^3
	V. parahaemolyticus[d]	107	12	2	5[e]	0	10^2	—
7 Frozen raw breaded shrimp and prawns	SPC[g]	64	1	3	5	3	10^6	10^7
	Faecal coliforms (MPN)	77	4	3	5	3	4	400
	Staphylococcus (ind.)[c]	114	4	3	5	3	10^3	2×10^3
	V. parahaemolyticus[d]	107	10	2	5[e]	0	10^2	—
8 Cooked picked crabmeat	SPC[g]	64	3	3	5	1	10^5	10^6
	Faecal coliforms (MPN)	77	6	3	5	1	4	400
	Staphylococcus (path.)[c]	114	9	3	5[e]	1	10^3	2×10^3
	V. parahaemolyticus[d]	107	12	2	5[e]	0	10^2	—

a This column refers to the pages in Thatcher and Clark (1968) where the methods are described. Use sample unit sizes recommended in the methods, except where otherwise indicated.
b Incubation temperature for SPC should be 25° C except for items 6 and 8 which should be incubated at 35° C.
c (ind.) refers to *Staphylococcus* as an indicator, (path.) as a toxinogenic pathogen.
d Testing for *V. parahaemolyticus* is recommended only for products from areas of high incidence of the organism (e.g., S.E. Asia and Japan) and hazardous products harvested during warm weather periods in other areas.
e Case 12 normally calls for $n = 15$, $c = 0$, case 10, for $n = 20$, $c = 0$, and case 9 for $n = 10$, $c = 1$ (see Table 11, p. 60). These are not practicable for routine analysis in most fish-testing laboratories.
f *Salmonella* testing and more stringent faecal coliform requirements are recommended only for fish from warm waters.
g Average SPC for shrimps in normal international trade are very high and SPC is of doubtful significance.

the choice of case (Table 17), derived from Table 11 (p. 60). Fresh products to be cooked have been assigned cases in the 'reduces concern' column of Table 11 (p. 60, cases 1, 4, 7, and 10), while cooked products, not likely to receive further heating, have been assigned cases in the 'may increase concern' column (cases 3, 6, 9, and 12). Precooked products likely to receive additional cooking in the home (e.g., fish sticks) have been given cases in the 'no change in concern' column (cases 2, 5, and 8).

For economic reasons, the number of sample units to be examined, n, is limited to five. Even five samples would tax the laboratory facilities available in many areas which import large quantities of fish. Probabilities of acceptance were chosen in relation to known commercial achievement and degree of hazard. In all instances, except for *Salmonella*, 3-class plans were considered to be highly advantageous.

Routine salmonellae testing is recommended only for warm-water fish raised by artificial culture or harvested from ponds, lakes, and rivers in highly populated areas with warm climates (Guelin, 1962; Gulasekharam *et al.*, 1956). Cold-water fish, such as the salmonids, which are cultured and harvested from waters with low ambient temperatures, present little or no risk of salmonellae infection and hence routine examination for salmonellae is not recommended. In the event of specific cause for concern, investigational *Salmonella* testing may be done. In such cases, the suspected product should be treated as case 12, using the plan $n = 20, c = 0$.

The cases considered applicable to fish products and the recommended sampling plans and microbiological limits are listed in Table 18. The tests recommended are standard plate count as a utility indicator, faecal coliforms, and/or coagulase-positive staphylococci as an indicator of post-harvesting contamination. In addition, the *Staphylococcus* test is used to indicate a potential for food-poisoning in cooked shrimp and crab meat, epidemiological evidence having substantiated a significant risk from such foods. It is further recommended that routine analysis for *V. parahaemolyticus* should be applied to products from areas where the incidence of the organism is high, e.g. Japan and South East Asia, and to shellfish products harvested in other areas during warm weather periods.

C SAMPLING PLANS

Tentative sampling plans are proposed for eight product categories on which a good deal of microbiological data is available (Table 18). Adequate data are still not available to permit the proposing of microbiological criteria for a number of other products in international trade which may present health hazards. Data on these products should be collected so that recommendations can be made. They include: (i) hot smoked fish (eels, chubs); (ii) cold smoked fish including kippers and bloaters; (iii) comminuted fish products (fish sausage and fish pastes); (iv) semi-preserved products; (v) sauces (e.g. Nuoc-Nam, salted and dried fish products); (vi) fish-egg products (caviar); (vii) squid and octopus; (viii) miscellaneous specialty products (e.g., sea urchin eggs); (ix) cultured warm-water fish and crustacea.

D SAMPLING PROCEDURES

For general directions on collecting and handling field samples and sample units, see Chapter 7. In routine analysis, collect five field samples (see Table 18) from each lot (or consignment) and in each case take at least twice as much food as will be required for the laboratory analyses (see Section B, page 83). Use 0.1% peptone water (Thatcher and Clark, 1968, p. 182) as the diluent for all tests; for fish and fish products, experience has shown this to be the best diluent.

(a) *Iced or chilled raw fish*

1 Whole fish in international trade is normally destined for further processing and is not usually tested bacteriologically. Where whole fish are to enter directly into the retail market, bacteriological tests are normally directed at obtaining estimates of bacterial populations on skin surface for large fish and total bacterial count/unit weight in the case of small fish. Field samples usually consist of single fish in the case of large fish and one or more fish in the case of small fish. Surface samples may consist of swabs taken over a prescribed surface area (e.g., 200 sq cm), or of an aseptically excised area of skin including a minimum of underlying tissues. Shake the swab or skin tissue thoroughly in 10 ml of peptone diluent and proceed with analyses as described in Thatcher and Clark (1968). Express results as counts per unit area (e.g., sq cm)

2 For chilled or iced fillets of small fish, take a whole fillet for each field sample; for large fish such as halibut, take a representative portion (e.g., three sections from different parts of the fillet). In the laboratory, weigh out a sample unit of 100 g from each field sample and proceed with blending, dilution, and analysis according to Thatcher and Clark (1968). The initial dilution in blending can range from 1:3 to 1:10.

3 For scallops in ice, collect one field sample from each of five containers. Weigh out 25 g sample units from each field sample and proceed with blending, dilution, and analysis as described in Thatcher and Clark (1968).

(b) *Frozen fish*

Frozen fin fish products are normally shipped as blocks of whole gutted fish, fillet blocks, comminuted fish blocks, or consumer packages of fillets or other prepared material.

1 Large fish such as salmon, tuna, swordfish, or halibut may be shipped as frozen whole (usually gutted) fish with or without heads. Sampling and bacteriological testing of such products is difficult and generally is not done. In certain cases imported products of this type are detained until examination of a sample from subsequent processing (e.g., steaking operation) has been shown to be satisfactory. Examination of the processed sample follows the methods given in (2) below.

2 Field samples of wholesale units of frozen fish may consist of individual fish, or entire blocks, cartons, or institutional packages; or they may be pieces cut, sawn, or drilled from these larger units (preferably during a primary processing operation in the receiving plant). In the laboratory, thaw field samples for three hours at room temperature (20–25° c) or overnight in a chilled room (2–5° c) and weigh out 100 g sample units for analysis. Whenever possible the sample unit should be blended with diluent before becoming completely defrosted. The temperature of the product should not rise above 5° c prior to analysis. Conduct blending, dilution, and analysis according to Thatcher and Clark (1968). The initial dilution can range from 1:3 to 1:10.

3 For sampling retail packages of frozen fish, follow instructions given in Section c, page 123. Again, use 100 g sample units.

4 Sample and analyse frozen raw or cooked shrimp, prawns, and lobster tails, and frozen raw breaded shrimp (items 5–7, Table 18) as described in (2) and (3) above.

(c) *Cooked and cold smoked fish*

Sample and analyse as described for raw (fresh or frozen) fish above.

9

Sampling plans for vegetables

Vegetables, both fresh and frozen, are becoming increasingly popular in international commerce. For example, there is a long-established and growing trade among the European nations; and an increasing importation of lettuce and other fresh vegetables from the United States into Europe, and of tomatoes from Mexico into the United States.

This chapter deals primarily with fresh; chilled or frozen (raw); blanched, frozen; and canned vegetable commodities. Formulated vegetables (e.g., precooked, frozen, and containing other ingredients such as sauces), as well as dried vegetables, are dealt with in other chapters.

A FRESH VEGETABLES

There have been numerous reports of disease outbreaks among humans as the result of consuming uncooked fresh vegetables. Most frequently reported are salmonellosis, bacillary dysentery, cholera, leptospirosis, infectious hepatitis, viral gastroenteritis, and amoebic dysentery. Less common diseases associated with the growing of vegetable crops under irrigation are brucellosis, tuberculosis, tularemia, swine erysipelas, coccidiosis, ascariasis, cysticercosis, fascioliasis, schistosomiasis, and hookworm and tapeworm infestations. Salmonellosis outbreaks, including typhoid and paratyphoid, have been associated with the consumption of watercress (Melick, 1917), lettuce (Kreuz, 1955), cabbage (Gaub, 1946), raw salad vegetables (Harmsen, 1953), and fruits (Cotrufo *et al.*, 1957). Heavy roundworm infestations have been reported in Europe and the Orient to result from the use of night soil in gardens. An outbreak of liver fluke infestation in England was associated with the consumption of wild watercress (Hardman *et al.*, 1970).

All such disease outbreaks associated with the consumption of raw vegetables are related to surface contamination of the vegetables by the

TABLE 19

Sampling plans and recommended microbiological limits for vegetable products

Product	Test	Method reference[a]	Case	Plan class	n	c	Limit per g	
							m	M
Fresh vegetables (to be consumed raw)	*E. coli*	77	5	3	5	2	10	10^2
	Salmonella	90	11	2	10	0	0	—
Chilled or frozen vegetables (to be consumed raw)	*E. coli*	77	5	3	5	2	10	10^2
	Salmonella	90	11	2	10	0	0	—
Fresh vegetables (to be cooked)	*Salmonella*	90	11	2	10	0	0	—
Blanched, frozen vegetables	SPC	64	4	3	5	3	10^5	10^6
	E. coli	77	5	3	5	2	10	10^2

a This column refers to the page numbers in Thatcher and Clark (1968) where the methods are described. Use sample unit sizes recommended in the methods.

infectious agent as the result of practices followed in the growing of the crops, or in their processing, packing, and shipment to market. None of the pathogens is known to be a plant pathogen or to grow on the vegetables. One report, however (Gayler *et al.*, 1955), demonstrates that salmonellae can grow on the cut surface of watermelon.

There is a growing market in which fresh vegetables are chopped, packaged, then chilled or frozen, and intended to be consumed uncooked in the form of salads. Such preparations are subject to the same degree of contamination as, if not more than, the conventional fresh vegetables. Fresh vegetables intended to be cooked prior to eating would be less hazardous, since all of the pathogens noted above are heat sensitive. Nevertheless, contaminated vegetables may serve as a source of pathogenic microorganisms for contaminating other foods.

Table 19 summarizes a tentative sampling plan for the examination of fresh vegetables. Because of the conditions under which vegetables are marketed, both locally and internationally, it is unlikely that such examinations could be conducted on a routine basis. It is therefore suggested that the sampling be conducted only on lots whose poor condition has brought them under suspicion, or are known to have been produced under substandard agricultural practices, or connected with an epidemiological survey.

It will be noted in Table 19 that there are no suggestions for a standard plate count or coliform count on fresh vegetables. It is highly unlikely that such determinations would have any significance for human health, because of the possible presence of localized areas containing high populations of plant pathogens which would interfere with such examinations. Detection specifically of *E. coli* or *Salmonella*, however, would indicate faecal contamination of the plant surfaces.

The control and maintenance of good agricultural practices in growing the crops, combined with acceptable hygienic methods in the harvesting, preparation, packing, and transport of the vegetables in international trade, should prove to be of considerably greater significance than bacteriological testing in disease control. Geldreich and Bordner (1971) point out the importance of low faecal coliform counts in irrigation waters, and the use of potable water to wash and freshen the harvested vegetables. Several codes of sanitary practice and good manufacturing procedure are available. These codes are designed to minimize the chances of contamination of fresh vegetables as they are processed and packed for shipment (e.g., Joint FAO/WHO Food Standards Programme, 1969; USFDA, 1969; Elliott and Mitchener, 1961, review of earlier codes). The need for clean containers for packing vegetables was amply demonstrated by detecting *Salmonella* contamination of produce packed in wooden crates formerly used for shipment of iced poultry (USDHEW, 1968).

B BLANCHED, FROZEN VEGETABLES

Prepackaged blanched frozen vegetables continue to be popular convenience foods and appear to some extent in international trade. The blanching, primarily designed to inactivate degradative enzymes in the vegetables, is sufficient to kill any of the aforementioned pathogens that may appear on the surfaces. Therefore, the bacteriological condition of such products is primarily a reflection of the sanitary practices employed after blanching. In some instances it may be difficult to prevent cross-contamination of coliforms into the blanched products. Standard plate counts and assessment of coliform population are suggested in Table 19 for evaluating their wholesomeness.

In a rather extensive survey conducted on blanched frozen vegetables, Splittstoesser and Segen (1970) failed to detect *Salmonella* contamination. It is therefore doubtful whether *Salmonella* tests on such products would prove to be of value. Generally, such vegetables have enjoyed a commendable safety record, probably due in part to the control procedures followed by the processors.

C CANNED VEGETABLES

Canned low-acid vegetables if inadequately processed pose a major threat to public health – botulism. Because of this, most canners follow rigidly the so-called 12-D concept in the retorting of low-acid vegetables, and, as a result, botulism from commercially canned vegetables has become rare. Safe processes for virtually all canned vegetables are now published (National Canners Association, 1966, 1971) and calculated on the basis of careful determinations of heat penetration for each commodity in different sizes of container. Nevertheless, the safe procedures may be neglected or human error may occur, thus creating a hazard.

Chapter 15 sets down principles applicable to canned foods in general.

Bacteriological examination is inappropriate for assessing the safety of canned low-acid vegetables because the stringencies required far surpass any laboratory technique now available. As for canned meats (Chapter 14), a statistical approach would be inapplicable, and therefore canned foods are not listed in Table 19.

Canned acid vegetables or fruits having a pH value of 4.6 or less do not present any microbiological problems related to public health. However, incubation tests may be appropriate where processing data are lacking (see Chapter 15). For low-acid vegetable products liable to 'flat sour' spoilage, the pH values on the contents of all cans to be torn down for seam examination should be determined (see Chapter 15).

In general, the processor should not be required to furnish records indicating results of incubation tests on sample cans of conventionally processed low-acid foods. Rather, it is far more important to request records pertaining to the thermal process as well as container integrity. If such records are not available, it is recommended that the controlling agency sample individual lots as outlined in Chapter 15 keeping in mind, of course, the limitations of any sampling or incubation procedure with respect to assessing the botulism hazard.

D SAMPLING PROCEDURES

The sampling of fresh, frozen, or processed vegetables presents no problems peculiar to these foods. Fresh vegetables may be sampled in accord with the principles discussed in Chapter 7, but because the numbers and dominant types of microorganisms can change rapidly in fresh vegetables at ambient temperatures, precautions for refrigeration during transport to the laboratory are essential, as described in Section (g), page 88. Frozen vegetables may be sampled and handled as described in Section C, page 123. For retail commodities, always collect unopened packages as field samples. For fresh vegetables in large containers (e.g., crates), collect field samples in sterile plastic bags; take only one sample from each container opened, using random procedures (Section C, page 11).

Canned vegetables, for which a 'botulinum cook' is essential for safety, are sampled as described for shelf-stable canned foods, in Chapter 15, Sections E and F. Microbiological examination is not normally required. Where processing data are lacking, visual inspection for swells and examination of seams may be indicated. A sequence of such procedures is listed in Table 26 (page 150).

10

Sampling plans for dried foods

A large number of dried foods – materials of animal, cereal, vegetable, and fruit origin – are moving increasingly in international commerce. While because of their low a_w these foods are usually shelf-stable at ambient temperatures, they may nonetheless contain various pathogens. The potential exists, therefore, for food poisoning when contaminated dried foods are eaten directly, or after rehydration, or if they contaminate other foods. Surveillance is obviously necessary in many situations; and it is particularly important when the food is destined for areas where refrigeration is inadequate, or when the food will be eaten by high-risk consumers (Table 8, p. 42).

A SAMPLING PLANS

Of the many different types of dried foods in international commerce, only those which experience has shown can present a microbiological hazard have been considered. As an initial step towards establishing appropriate cases (Table 6, p. 33), these selected dried foods have been classified as finished products to be sold to the consumer (Table 20) or as ingredients (Table 21) to be used in the manufacture of other food products. These foods have been subdivided further according to their origin, i.e. animal, cereal, vegetable, and fruit. This approach seems helpful in that foods within each final category tend to represent similar health or utility risks, and hence to be subject to the same or similar sampling plans. Dried milk foods are dealt with in the chapter on milk and milk products (page 127).

Tables 20 and 21 give sampling plans and limits of acceptance for those tests considered appropriate for routine analysis. These are suggested for application where there is reason to believe from experience

TABLE 20

Sampling plans and recommended microbiological limits for dried foods: finished products known to present microbiological hazards

Product	Test	Method reference[a]	Case	Plan class	n	c	Limit per g	
							m	M
I ANIMAL ORIGIN								
1 Beverage, meat	*C. perfringens*	127	8	3	5	1	10^2	10^4
2 Dietetic foods	SPC	64	2	3	5	2	10^4	10^6
	Coliforms or Enterobacteriaceae	69 82	5	3	5	2	10	10^3
	Salmonella[b]	90	11	2	10	0	0	—
3 Egg products	SPC[c]	64	2	3	5	2	10^4	10^6
	Coliforms or Enterobacteriaceae[c]	69 82	5	3	5	2	10	10^3
	Salmonella[b]	90	11	2	10	0	0	—
4 Proteins	SPC	64	2	3	5	2	10^4	10^6
	E. coli	77	5	3	5	2	<3[d]	10
	C. perfringens	127	8	3	5	1	10^2	10^4
	Salmonella[b]	90	11	2	10	0	0	—
5 Soups not to be cooked	SPC	64	2	3	5	1	10^4	10^6
	Coliforms or Enterobacteriaceae	69 82	5	3	5	2	10	10^3
	C. perfringens	127	8	3	5	1	10^2	10^4
	Salmonella[b]	90	11	2	10	0	0	—

TABLE 20 (Continued)

Product	Test	Method reference[a]	Case	Plan class	n	c	Limit per g m	Limit per g M
6 Special dietary foods[e]	SPC	64	3	3	5	1	10^4	10^6
	E. coli	77	5	3	5	2	$<3^d$	10
	Staph. aureus	114	9	3	10	1	10	10^2
	B. cereus	138	9	3	10	1	10^2	10^4
	C. perfringens	127	9	3	10	1	10^2	10^3
	Salmonella[b]	90	15	2	60	0	0	—
II CEREAL FOODS								
1 Cake mixes with high egg content	*Salmonella*[b]	90	11	2	10	0	0	—
2 Dietetic foods	See product I, 2 above, this table (dietetic foods animal origin)							
3 Pasta with egg	SPC	64	2	3	5	2	10^4	10^6
	Staph. aureus	114	8	3	5	1	10	10^3
	Salmonella[b]	90	11	2	10	0	0	—
4 Pudding with egg	SPC	64	2	3	5	2	10^4	10^6
	E. coli	77	5	3	5	2	$<3^d$	10
	Staph. aureus	114	8	3	5	1	10	10^3
	Salmonella[b]	90	11	2	10	0	0	—
5 Snacks dusted with flavouring	*Salmonella*[b]	90	11	2	10	0	0	—
6 Special dietary food	See product I, 6 above, this table (special dietary foods animal origin)							
III FRUITS								
1 Dried fruits (dates, figs, etc.)	*E. coli*	77	5	3	5	2	$<3^d$	10

TABLE 20 (Concluded)

Product	Test	Method reference[a]	Case	Plan class	n	c	Limit per g	
							m	M
IV VEGETABLE FOODS								
1 Dietetic foods (low calorie)	See product I, 2 above, this table (dietetic foods of animal origin)							
2 Frostings with eggs, not to be cooked	*Salmonella*[b]	90	11	2	10	0	0	—
3 Nut butters	*Salmonella*[b]	90	11	2	10	0	0	—
4 Potatoes	*B. cereus*	138	8	3	5	1	10^3	10^5
	C. perfringens	127	8	3	5	1	10^2	10^4
	Staph. aureus	114	8	3	5	1	10^2	10^4
5 Proteins (meat and poultry analogues)	SPC	64	2	3	5	2	10^4	10^6
	C. perfringens	127	8	3	5	1	10^2	10^4
6 Soups not to be cooked	SPC	64	2	3	5	2	10^4	10^6
	C. perfringens	127	8	3	5	1	10^2	10^4
	Staph. aureus	114	8	3	5	1	10^2	10^4
	Salmonella[b]	90	11	2	10	0	0	—
7 Vegetables analogues (e.g., imitation cherries)	SPC	64	2	3	5	2	10^4	10^6
	E. coli	77	5	3	5	2	< 3[d]	10

a This column refers to the page numbers in Thatcher and Clark (1968) where the methods are described. Use sample sizes recommended in these methods.
b All *Salmonella* tests are subject to compositing of the sample units (Silliker and Gabis, 1973; Gabis and Silliker, 1974).
c These plans do not apply to egg albumin desugared by bacterial fermentation.
d '<' 3 means no positive tube in the standard 3-tube MPN technique.
e Food to be eaten by high risk category of consumers, e.g., foods for infants and the aged, relief foods, etc.

TABLE 21

Sampling plans and recommended microbiological limits for dried foods: ingredients known to present microbiological hazards

Product	Test	Method reference[a]	Case	Plan class	n	c	Limit per g m	Limit per g M
I ANIMAL ORIGIN								
1 Dyes	*Salmonella*[b]	90	11	2	10	0	0	—
2 Egg products[c]	SPC	64	2	3	5	2	10^4	10^6
	Salmonella[b]	90	11	2	10	0	0	—
3 Enzymes	*Salmonella*[b]	90	11	2	10	0	0	—
4 Meats or components including gelatin and fish protein concentrate	*C. perfringens*	127	8	3	5	1	10^2	10^4
	Staph. aureus	114	8	3	5	1	10^2	10^4
	Salmonella[b]	90	11	2	10	0	0	—
5 Sea food	SPC	64	2	3	5	2	10^5	10^7
	E. coli	77	5	3	5	2	<3[d]	10
	Staph. aureus	114	8	3	5	1	10^2	10^4
II CEREAL ORIGIN								
1 Cereal by-products (bran, flours, etc.)	Moulds[e]		2	3	5	2	10^2	10^4
	Spores of thermophilic bacteria[f]		2	3	5	2		
	Spores of rope-forming bacteria[g]		2	3	5	2	10^2	10^4
	B. cereus	138	8	3	5	1	10^3	10^5
	C. perfringens	127	8	3	5	1	10^2	10^4
III FRUIT BASED								
1 Fruits, sun dried	Osmophilic yeasts[h]		2	3	5	2	10	10^3
	Moulds[e]		2	3	5	2	10^2	10^4
	E. coli	77	5	3	5	2	<3[d]	10
IV VEGETABLE ORIGIN								
1 Cocoa	SPC	64	2	3	5	2	10^4	10^6
	Moulds[e]		2	3	5	2	10^2	10^4

TABLE 21 (Concluded)

Product	Test	Method reference[a]	Case	Plan class	n	c	Limit per g m	Limit per g M
2 Coconut	Moulds[e]		2	3	5	2	10^2	10^4
	Coliforms or Enterobacteriaceae	77, 82	5	3	5	2	10	10^3
	Salmonella[b]	90	15	2	60	0	0	—
3 Dyes	SPC	64	2	3	5	2	10^4	10^6
4 Enzymes	*E. coli*	77	5	3	5	2	<3[d]	10
5 Gums	SPC	64	2	3	5	2	10^4	10^6
	Coliforms or Enterobacteriaceae	69, 82	5	3	5	2	10	10^3
6 Nuts	Moulds[e]		2	3	5	2	10^2	10^4
	E. coli	77	5	3	5	2	<3[d]	10
7 Spices	SPC	64	2	3	5	2	10^4	10^6
	Moulds[e]		2	3	5	2	10^2	10^4
	E. coli	77	5	3	5	2	10	10^3
8 Yeasts	*E. coli*	77	5	3	5	2	<3[d]	10
	Salmonella[b]	90	11	2	10	0	0	—
9 Vegetables	*E. coli*	77	5	3	5	2	<3[d]	10^2
	Salmonella[b]	90	11	2	10	0	0	—

a This column refers to the page numbers in Thatcher and Clark (1968) where the methods are described. Use sample sizes recommended in these methods.
b All *Salmonella* tests are subject to compositing of the sample units (Silliker and Gabis, 1973; Gabis and Silliker, 1974).
c These plans do not apply to egg albumin desugared by bacterial fermentation.
d '<' 3 means no positive tube in the standard 3-tube MPN technique.
e The method of Mossel *et al.* (1970) is suggested.
f The method of the National Canners' Association of the US (National Canners Association, 1968) is recommended.
g The method recommended by Sharf *et al.* (1964) is suggested.
h The method of Mossel & Bax (1967) is suggested.

with the product, or knowledge of its history, that no unusual hazard is likely to exist. For situations in which investigational examination is required (see Section N, page 63), additional sample units should be examined and additional tests may be required as well. For *Salmonella* the number of sample units recommended for investigational sampling are given in Table 13 (page 69); compositing of sample units will markedly reduce the work and expense involved (see page 70). The additional tests required will depend on the likelihood (based on experience) that the food could contain other pathogens; or, in the case of a disease outbreak, on the nature of the illness. For example, when information on production practices is inadequate, dried dietetic foods (Table 20, item I,2) should be examined for *B. cereus*, *C. perfringens*, and *Staphylococcus* as well as being given the routine tests listed. Similarly, egg products (Table 20, item I,3) may require a test for *Staphylococcus*; and soups to be cooked (Table 20, item I,5) may require tests for *B. cereus* and *Staphylococcus*. Additionally, some dried foods not routinely examined microbiologically may, in unusual circumstances, present a health risk. In these cases an investigational examination is indicated; for example, soft candy and chocolate occasionally contain enteropathogenic *E. coli* and *Salmonella*.

Dietetic foods of animal and of vegetable origins have been given the same sampling plans because of the likelihood that each type may contain ingredients from both sources.

A number of dried foods, e.g. nuts and cereal products, may present problems resulting from the presence of mycotoxins. The most effective mode of quality control in this instance is control at the source (see Chapter 6). When this cannot be achieved, investigational testing for the possible causative organism(s) and direct examination for mycotoxins is indicated.

In many instances, tests for coliforms and Enterobacteriaceae have been listed jointly. The reason for this is that, whereas the coliform test is customarily used in most areas of the world to indicate faecal contamination, the Enterobacteriaceae test is being used increasingly for this purpose in some countries, to the extent that it is being incorporated into current legislation.

Most tests for products in Tables 20 and 21 have been assigned cases in the 'no change in concern' column of Table 11 (page 60; cases 2, 5, 8, and 11). Ordinarily, increase in concern after rehydration would not be indicated provided the reconstituted product was used or consumed quickly or stored in a way to prevent proliferation of the incident microorganisms. Where microbial growth might occur, plan stringency could be increased by using cases 3, 6, 9, 12, and 15. This has been done in Table 20 for special dietary foods (item I,6) which are usually eaten by high-risk consumers (Table 8, page 42).

The values for *m* and *M*, as expressed in Tables 20 and 21, reflect a consensus of opinions based on knowledge of large numbers of analytical results derived from private and government laboratories and from the in-plant control testing of some large-scale manufacturers. The values represent commercially attainable quality under conditions of good manufacturing practice, the upper limit, *M*, being the value often used in purchasing specifications. The *c* values essentially follow those listed in Table 11 (page 60), which provide a guide for *c* values related to an increasing plan stringency in accord with increasing degree of concern (cases 1–15).

Generally, the sampling plans for *Salmonella* given in this chapter refer to serotypes appropriate to cases 10 to 12 (see Table 7, page 34). However, for the examination of shredded, desiccated coconut (Table 21, item IV,2), a case 15 plan has been recommended, because this product is commonly consumed without cooking and has been implicated in outbreaks of typhoid and paratyphoid. Also a case 15 plan has been recommended for special dietary foods (Table 20, item I,6), because these will be eaten by high-risk consumers (see Table 8, page 42).

B SAMPLING PROCEDURES

(a) *Collecting the sample*

1 Generally dried foods do not present any particular sampling problem. Most are homogeneous and composed of free-flowing small particles. Follow the general directions given in Chapter 7.

2 It is preferable to sample dried foods, in bulk, in ship compartments or in rail cars, during loading or unloading. In this way sample units can be obtained from widely different areas of the lot. If this procedure is not possible then use a sterile probe of the type commonly used for collecting sample units of grain or flour. Probes up to 75 cm long are available. Collect only one field sample unit per insertion of the probe. Use a separate sterile probe for each field sample unit or sterilize the probe between collection of units. See Section C(b), page 85 for directions on how to sterilize sampling instruments.

3 For foods contained in bags or barrels, use the precautions described in Section C (c), page 86 for opening the containers. If possible, use a probe (about 30 cm in length) similar to that described in (2) above,

inserting it diagonally. Use a separate sterile probe for each field sample unit or sterilize the probe between collections of units. Choose bags or barrels to be sampled at random and take only one sample per container.

4 For foods packaged in bulk in cartons, proceed as described for bags in (3) above.

5 For cartons of packages, randomly select one unopened package as the sample unit.

6 Sample dried fruits in cases or cartons using a trier, such as that used for sampling cheese (see Section C (d), page 135).

(b) *Preparing sample units for analysis*

1 For most dried foods, proceed as described in Section C (i), page 90 and in Thatcher and Clark (1968, p.60) for preparation of the food homogenate.

2 For the *Salmonella* test, sample units can be composited according to Silliker and Gabis (1973) (see page 70).

3 For gums, prepare the homogenate at 1/100 concentration instead of the usual 1/10. A 1/10 dilution will be too viscous to pipette. If the greater dilution means that certain microorganisms may be missed, use larger than usual aliquots.

4 A high fat content often complicates the examination of dried foods. To remove fat, centrifuge the food homogenate (1:10 suspension) at 3000 rpm for 10 min. Discard the liquid phase on top and resuspend the residue in the same volume of diluent used in the first dispersion. Blend in a blender for 1 min and proceed with the test analysis according to the procedures given in Thatcher and Clark (1968).

5 Occasionally, it may be necessary to analyse separately the unprocessed spice component of dusted products (e.g., paprika chips). In such cases aseptically scrape the spice free from the baked part, using a sterile scalpel, and analyse the scrapings.

6 For low-acid products such as beverage powders, adjust the pH of the food homogenate to 6.8 – 7.0 or alternatively use a buffered diluent when preparing the homogenate (Thatcher and Clark, 1968, p. 60). For determination of enteric organisms such as *E. coli*, incubate the neutralized homogenate at 35 to 37° C for about three hours to allow the acid-injured cells to resuscitate (Mossel and Ratto, 1970; Roth and Keenan, 1971).

11

Sampling plans for frozen foods

Frozen foods consisting of one commodity only such as meat, poultry, fish, vegetables, or a milk product are considered in other sections of this report. Only frozen egg products, formulated pre-cooked entrees (dishes or meals), vegetables in sauces, and ready-to-eat desserts (except ice cream, ice milk, and related desserts) are included in this chapter. Although many kinds of food may be frozen commercially, the above selections are typical of the major categories that enter international trade. Many snack foods and ethnic dishes are similar to the more common pre-cooked pot pies, meat with gravy, or complete meals. Other items, such as biscuit and bread dough, have been omitted, either because insufficient information was at hand to make judgments about their microbial quality, or because they seemed to have limited potential for the transmission of disease.

The excellent record of the frozen food industry in making safe wholesome products for many years is frequently given as a reason for not imposing microbiological standards on these products. In large measure, this record is a result of self-imposed good manufacturing practices and quality control programs. There is substantial evidence (AFDOUS, 1969; Surkiewicz, 1966; Leininger *et al.*, 1971; Takács *et al.*, 1968, 1969; Canadian Department of Agriculture, 1971) that firms differ considerably in this respect, and the poorly controlled operations could be potential sources of massive food-borne disease outbreaks. In fact, a single production lot, if contaminated with an infectious or toxic agent, could expose thousands of consumers to the risk of disease. In order to protect the consumer's health and industry's reputation by minimizing this potential hazard, sampling plans and recommended microbiological limits have been listed below. The values are based on professional judgment and on limited data available from both published and unpub-

TABLE 22

Sampling plans and recommended microbiological limits for selected frozen foods

Product	Test	Method reference[a]	Case	Plan class	n	c	Limit per g m	Limit per g M
Entrees, pre-cooked, and vegetables in sauce	SPC	64	5	3	5	2	10^5	10^6
	Coliforms	73	5	3	5	2	10^2	10^4
	E. coli	77	5	3	5	2	<3[b]	10^2
Desserts	SPC	64	5	3	5	2	10^4	10^6
	Coliforms	73	5	3	5	2	10^2	10^4
	E. coli	77	5	3	5	2	<3[b]	10^2
	Staphylococci	114	5	3	5	2	10	10^3
Whole egg and yolk products (pasteurized)	Direct microscopic count[c]	—	1	3	5	3	5×10^5	5×10^6
	SPC	64	4	3	5	3	10^4	10^6
	Salmonella	90	10	2	5	0	0	—
Egg whites (desugared, pasteurized)	SPC	64	5	3	5	2	10^4	10^6
	Salmonella	90	11	2	10	0	0	—
Egg whites (not desugared, pasteurized)	Direct microscopic count[c]	—	2	3	5	2	5×10^5	5×10^6
	SPC	64	5	3	5	2	10^4	10^6
	Salmonella	90	11	2	10	0	0	—

a This column refers to pages in Thatcher and Clark (1968) where the methods are described. Use sample unit sizes recommended in the methods.

b '<' 3 means no positive tube in the standard 3-tube MPN technique.

c Refer to Horwitz (1970) for method for Direct Microscopic Count.

lished sources (AFDOUS, 1969; Surkiewicz, 1966; Leininger *et al.*, 1971; Takacs *et al.*, 1968, 1969; Canadian Department of Agriculture, 1971; Hillig *et al.*, 1960; Steinhauer *et al.*, 1967; Reagan *et al.*, 1971).

A FROZEN PRECOOKED ENTREES, VEGETABLES IN SAUCES, AND CREAM-TYPE OR CUSTARD-FILLED DESSERTS

Table 22 summarizes recommended microbiological limits for SPC, coliform group, and *E. coli* for frozen pre-cooked entrees, vegetables in sauces, and desserts. It also includes limits for coagulase-positive staphylococci in cream-type or custard-filled desserts. The recommendations are based on limited data (AFDOUS, 1969; Surkiewicz, 1966; Leininger *et al.*, 1971; Takacs *et al.*, 1968; Canadian Department of Agriculture, 1971); however, these limits are attainable by good manufacturing practices (GMP) in the frozen food industry. AFDOUS made a survey of beef pies and chicken pies in 1966 and 1967 (AFDOUS, 1969) and the Meat and Poultry Inspection Program of USDA made a national survey of firms producing cooked meat and gravy, and cooked meat patties in 1970–1. The data from the latter survey were measured against the recommended limits for 'entrees, pre-cooked.' There were some violations, but each was corrected quickly to bring the processing techniques to an acceptable level.

Tests for entrees and desserts in Table 22 have been assigned cases in the 'no change in concern' column of Table 11 (page 60; cases 2 and 5). Ordinarily, increase in concern after thawing would not be indicated provided the thawed product was consumed without delay or was refrigerated (<5°C) until eaten. Where microbial growth might occur in the thawed product, thus leading to a health or spoilage risk, plan stringency could be increased by using cases in the 'increases concern' column of Table 11.

The indicated values of *m* and *M* are easily attainable in plants employing GMP, except when raw products such as cheese, parsley, or celery are added after cooking. The criteria shown do not apply to this latter class of products.

For the SPC, the values selected for *m* are slightly higher than are attainable by good manufacturing practice. *M* is the level at which some products begin to deteriorate organoleptically (reviewed by Elliott and Michener, 1961).

For coliform tests, the value chosen for *m* (100/gm) is attainable only when these products are produced under GMP. Values exceeding *M*

(10,000/gm) indicate that either excessive growth of the initial contaminants has been allowed, or that the contamination has been substantial.

The recommended limits for *E. coli* were set lower than for coliforms because of the potentially more serious nature of faecal contamination indicated by the presence of this species.

The sampling plan for *Staphylococcus* is recommended only for frozen desserts. It is primarily an indicator of unhygienic practices, but it may also have some value as a 'hazard' test. Normally these products should be free from *Staphylococcus*, but because of the ubiquitous nature of this organism, the value for m was selected to allow for the low levels of unavoidable contamination that occur occasionally. M was chosen as that value resulting from growth in the product or the introduction of an undesirably large amount of contamination. It may also permit only brief holding of the product after thawing because of the risk of the development of enterotoxin.

The sampling plans noted above are recommended for routine testing of products known to be produced under good manufacturing practice. Where the past history and source of the product are unknown or there is reason to question its microbiological quality, additional and perhaps more stringent sampling plans will be needed. For example, if epidemiological evidence were to suggest the possible presence of staphylococcal enterotoxin or *Salmonella* contamination, intensive investigative sampling would be indicated. A sampling plan for *Salmonella* is suggested in Section B, page 67.

B FROZEN EGGS

The microbiological tests employed are the direct microscopic count (DMC), the standard plate count (SPC), and examination for *Salmonella*. The DMC is used to detect the presence of decomposition, either from the inclusion of spoiled individual shell eggs, or from time-temperature abuse of the product. For this purpose, microbiological results are often used in conjunction with organoleptic tests and quantitative determinations of selected organic acids (Hillig *et al.*, 1960; Steinhauer *et al.*, 1967; Reagan *et al.*, 1971). A DMC exceeding approximately 5×10^6/ml indicates decomposition: thus, this figure denotes M. When DMCs are less than 300,000/g, their accuracy is questionable. The choice of m as 5×10^5 is related to this fact.

Tests for whole-egg and yolk products have been assigned cases in the 'reduces concern' column of Table 11 (page 60; cases 1, 4, and 10) because these products are almost always cooked prior to consumption. Egg whites, however, are very often not cooked prior to consumption or, if cooked, receive insufficient heating to kill *Salmonella*. These products, therefore, have been assigned cases in the 'no change in concern' column of Table 11 (cases 2, 5, and 11). Where microbial growth might occur in the thawed product and lead to a health or spoilage risk, plan stringency could be increased by using cases in the 'increases concern' column of Table 11 (cases 3, 6, and 12).

Standard plate count is used to check the adequacy of the pasteurization and is therefore an indicator test (case 4 or 5). A low count does not constitute a guarantee, but a high count indicates either insufficient pasteurization or subsequent abuse of the pasteurized product. SPC values of m and M were chosen from data submitted by Steinhauer *et al.* (1967) and the Canadian Department of Agriculture (1971).

The alpha-amylase test is frequently used, especially in the United Kingdom, as a test for underpasteurization. However, it cannot be applied generally, because some effective pasteurization procedures employed in the United States do not completely inactivate the enzyme.

All frozen eggs should be tested for salmonellae, because historically they have been a major source of human salmonellosis. The recommended plan is shown in Table 22 (see also Section B, page 67).

Some unpasteurized egg products are shipped between countries, and represent a serious hazard as vehicles for the dissemination of salmonellae. The international shipment of unpasteurized egg products should be discouraged, and their sale prohibited whenever legislation makes this possible. No recommended limits have been proposed for these products.

C SAMPLING PROCEDURES

The following steps are recommended for sampling, transporting, and obtaining portions of frozen foods for microbiological examination:

(a) *Frozen foods in retail packages*

1 Identify the lot to be sampled and divide it into five approximately equal parts, if possible.
2 Remove one package of the product from each part. If the packages are in shipping containers, take the packages from separate containers.
3 Quickly label each package to identify the lot, part, unit, number, etc. Avoid defrosting.

4 Place all packages in an insulated sample container with sufficient dry ice (solid carbon dioxide) to keep the samples frozen during transportation to the laboratory.
5 On arrival at the laboratory, examine the packages for proper labelling and evidence of thawing. Do not examine samples that have melted during transportation.
6 If the sample cannot be examined promptly, store the frozen packages in a freezer that maintains a maximum temperature of –10°c.
7 Aseptically remove representative portions from the frozen field sample with the aid of a sterile drill, chisel, or other device; or place the entire contents of a package into a sterile blender for mixing and then remove known amounts for testing. Follow directions given in Thatcher and Clark (1968, p. 60) for blending and diluting the food homogenate.
8 If necessary, 'temper' the frozen product by permitting it to thaw slightly prior to preparing the homogenate, but not to the point where all the ice is melted.
9 Analyse individual components of whole meals separately. Use the highest bacterial levels found in individual components to determine the acceptability of the product.
10 Retain reserve portions of field samples in the freezer.

(b) *Frozen eggs*

1 Identify the lot to be sampled and randomly select ten units for examination.
2 Examine, in accordance with the methods prescribed for frozen eggs in 'Official Methods of the Association of Official Analytical Chemists' 11th edition (Horwitz, 1970), section 41.003(b), 41.004(b) and 41.007(b), as follows:

Equipment. Electric (high speed) or hand drill with 40 × 2.5 cm auger; hammer and steel strip 30 × 5 × 0.5 cm, or other tool for opening cans; tablespoon; hatchet or chisel; precooled sterile containers (screw-cap jars or friction-top cans); alcohol lamp or other burner; cotton; clean cloth or towel; and water pail.

Sampling. Remove top layer of egg with sterilized hatchet or chisel. Drill 3 cores from top to bottom of container: first core in centre, second core midway between centre and periphery, and third core near edge of container. Transfer drillings from container with a sterile spoon to a prechilled sample container. Examine product organoleptically by smelling at opening of fourth drill-hole made after removal of bacteriological sample. (Heat produced by electric drill intensifies odour of egg material, thus facilitating organoleptic examination.) Record odours as normal, abnormal, reject, or musty. Place samples on solid CO_2 immediately so that drillings do not thaw.

Preparation of sample. Thaw frozen egg material as rapidly as possible to prevent increase in number of microorganisms present and at temperature low enough to prevent destruction of the microorganisms ($\leqslant$10°c for $\geqslant$15 min). (Frequent rotary shaking of sample container aids in thawing frozen material. Thawing temperature may be maintained by use of water bath or bacteriological incubator.)

D PROCEDURE FOR DIRECT MICROSCOPIC COUNTS

The following method is that described in 'Official Methods of the Association of Official Analytical Chemists' 11th edition (Horwitz, 1970), section 41.012.

Preparation of North aniline oil-methylene blue stain. Mix 3.0 ml aniline oil with 10.0 ml alcohol, and slowly add 1.5 ml HCl with constant agitation. Add 30.0 ml saturated alcohol methylene blue solution, dilute to 100.0 ml with water, and filter.

Procedure for liquid and frozen eggs. Place 0.01 ml undiluted egg material on clean, dry microscopic slide and spread over area of 2 sq cm (circular area with diameter of 1.6 cm suggested). Let film preparation dry on level surface at 35–40°c. Immerse in xylene $\leqslant$ 1 min; then immerse in alcohol $\leqslant$ 1 min. Stain $\geqslant$45 sec in North aniline oil-methylene blue stain (10–20 min preferred; exposure up to 2 hr does not overstain). Wash slide by repeated immersions in H_2O and dry thoroughly before examination. Perform subsequent operations and observe precautions as described in 'Standard Methods for Examination of Dairy Products,' (APHA, 1967). Express final result as a number of bacteria/g egg material (double microscopic factor, since 2 sq cm area is used).

12

Sampling plans for milk and milk products

In so far as feasible control techniques are concerned, milk and milk products comprise two broad categories: (i) milk *per se* and other rapidly consumed perishable dairy products such as cream, buttermilk and other fermented milks, flavoured milk or skim milk drinks, fresh cheese, and ice cream; (ii) processed commodities such as powdered or canned milk, ripened cheese, butter, powdered whey, lactose, sterilized or UHT milk, and ice cream. It appears from (i) and (ii) above that one product, ice cream, can be placed in either of the two categories. This will be discussed in Section B, below.

A PASTEURIZED FLUID MILK AND OTHER HIGHLY PERISHABLE PRODUCTS

The perishable products of this category cannot be subject to control on an a priori basis, using analysis of the end product, because the products will undoubtedly have been widely distributed within the retail trade and probably consumed before the microbiological examinations could have been completed. Only an a posteriori basis for control is practicable. Hence, the control methods commonly used are based on periodic analyses from which control authorities decide whether or not to accept a product in the future. The judgment is based upon the performance record of a specific producer or processor over a defined period of time, for example, upon the proportion of a specified number of analyses which complied with the standard. (See, for instance, the '3 out of 5' rate for compliance. U.S. Milk Ordinance and Code (USDHEW, 1965))

The health problems to be anticipated from pasteurized fluid milk and its allied products are likely to be local. Indeed, local public health authorities normally exercise control of milk as an important part of their responsibilities. Since the recommendation of microbiological limits for international trade in these products was therefore considered to be of low priority, sampling plans are not proposed. Present needs are probably best met by the application of mutually acceptable agreements between pairs of exporting-importing countries. The reports of the joint FAO/WHO Expert Committee on Milk Hygiene (Joint FAO/WHO 1957, 1960, 1970) provide valuable instruction in milk control procedures and disease hazards. Quality codes and criteria are described in publications of the International Dairy Federation (IDF, 1958a, 1958b) and the American Public Health Association (APHA, 1971); and in the Milk Ordinance and Code of the United States Public Health Service (US-DHEW, 1965).

The microbiological tests most often used as legal tests are the Standard Plate Count (SPC), the Coliform test, and the Reductase test (Methylene Blue or Resazurin tests). In addition, the Phosphatase test is in use worldwide because failure to detect the phosphatase enzyme provides assurance that the test milk has been heated to a temperature and during a period of time sufficient to kill pathogens likely to be present in raw milk. Moreover it allows detection of the presence of raw milk, which may, by accident, faulty procedure, or fraud have been introduced into the pasteurized milk.

B PROCESSED DAIRY PRODUCTS WITH EXTENSIVE SHELF-LIFE

The milk products currently of concern in international commerce are dried milk, sterilized and UHT milk, and cheese. Unlike fluid milk, these products are subject to test before the food is consumed and hence they fall into the category with which this book is chiefly concerned.

(a) *Dried milk*

Several 'grades' of dried milk powder occur in commerce. The recommended microbiological limits proposed here are basic to all dried milk. Use of more exacting criteria for grading purposes is a national prerogative.

The various cases for concern (see Table 6, page 33; and Table 11, page 60), identified as applicable to dried milk, are cases 5, 8, 9, and 12. Case 5 refers to an indicator measurement for which the aerobic plate count and coliforms are recommended in association with 3-class plans (see Table 11). Coliform organisms in dried milk may die out during

TABLE 23

Sampling plans and recommended microbiological limits for dairy products

Product	Test	Method reference[a]	Case	Plan class	n	c	Limit per g m	Limit per g M
Dried milk	SPC	64	5	3	5	2	5×10^4	5×10^5
	Coliforms	69	5	3	5	2	<3[b]	10^2
	Staph. aureus	114	8	3	5	1	10	10^2
Dried milk (for relief work)	*Staph. aureus*	114	9	3	10	1	10	10^2
	Salmonella	90	12	2	60[c]	0	0	—
Sterilized and UHT milk	SPC after pre-incubation	IDF (1971)	Special (see p. 129)	2	5	0	10	—
Ice cream (simple, i.e., without added ingredients)	SPC	64	5	3	5	2	10^4	2.5×10^5
	Coliforms[d]	69	5	3	5	2	10	10^3
	Staph. aureus	114	8	3	5	1	10	10^2
	Salmonella[e]	90	11	2	10	0	0	—
Ice cream (complex, i.e., with added ingredients)	SPC	64	5	3	5	2	2.5×10^4	2.5×10^5
	Coliforms[d]	69	5	3	5	2	10^2	10^3
	Staph. aureus	114	8	3	5	1	10	10^2
	Salmonella	90	11	2	10	0	0	—
Cheese: 'hard' and 'semi-hard' types[f]	*Staph. aureus*	114	8	3	5	1	10^3	10^4

a This column refers to page numbers in Thatcher and Clark (1968) where the methods are described. Use sample unit sizes recommended in the methods, except where otherwise indicated.
b '<' 3 means no positive tube in the standard 3-tube MPN technique.
c The 60 × 25 g unit samples may be aggregated into sample units of 3 × 500 g or 15 × 100 g (Silliker and Gabis, 1973; Gabis and Silliker, 1974).
d Because of the presence of sucrose in ice cream, coliform-like colonies on Violet Red Bile Agar should be confirmed.
e Investigational sampling only
f In addition, such cheese should be aged for 60 days at not less than 4.4° C (40° F) when made from unpasteurized or unheated milk or curd (see p. 132).

storage. They ought, nevertheless, to be absent, or present only in low numbers in dried milk, even when this dried milk has not yet aged. Thus, absence of coliforms is necessary but is not, of itself, sufficient to qualify a product as acceptable.

Coagulase-positive staphylococci are a potential hazard in dried milk, but normal conditions of use do not normally increase the hazard. Case 8 would therefore apply. If, however, the reconstituted milk prepared by adding a suitable amount of water to the dried milk were to be exposed to times and temperatures sufficient to permit growth of staphylococci, the hazard could be increased by the production of enterotoxin. Such conditions could apply if dried milk, after being reconstituted, were subjected to abuse during mass feeding operations such as those in hospitals, institutions, or in relief or emergency conditions. Case 9 applies in such instances, and international organizations administering aid programs should require that consignments of dried milk be tested according to the sampling plan recommended for case 9. As soon as sufficient supplies of specific enterotoxin antisera are available, direct examination for enterotoxin (Casman and Bennett, 1965) is desirable to investigate abuse that may have occurred.

Case 12, which indicates a serious hazard, applies because of the risk of outbreaks of salmonellosis arising from contaminated dried milk, when it is used for young children or any other susceptible population group. In such instances, but only in such instances, should samples be tested for salmonellae (see Section B, page 67, Section E, page 70, and Table 13, page 69) unless there is a need for investigational sampling.

The values chosen for m and M are listed in Table 23. Those for m are based on levels obtainable under good commercial processing. Most M values are kept low, first because milk and most milk products are largely consumed by infants, children, and other fragile groups; and second because milk and milk products are in many circumstances particularly favourable culture media for growth of common and pathogenic bacteria and develop hazards unusually rapidly.

(b) *Sterilized milk and aseptically filled* UHT *milk*

These products should comply with the concept of 'commercial sterility' (often referred to as 'appertisation'). Although they have some characteristics in common with other 'commercially sterile' canned milk products, such as evaporated milk, they also have some distinctive characteristics which influence control needs. The composition of sterilized or UHT milk is the same as that for normal fresh milk: the milk is not concentrated to any appreciable extent prior to heat treatment; the container may be a metal can, a glass bottle closed with a metal cap, a reinforced carton, or a plastic holder.

In the case of sterilized milk, any surviving organisms are usually thermophilic spore formers (detected by incubating the milk in its original container at 55°C) (Burton *et al.*, 1965). A contaminating flora will also be present if, at some stage after the sterilizing heat treatment, extraneous bacteria enter the product, usually the result of some fault in the closure of the container.

In the case of UHT *milk*, the milk is usually sterile immediately after heating, but potential opportunities for contamination are provided by (a) the filling operation, if it is not fully aseptic, and (b) leakage, particularly of cartons and plastic containers, at some stage during transportation or handling. In either instance ((a) or (b)), the expected bacterial contaminants would probably consist of a variety of species besides the thermophiles which may survive the heat treatment given to sterilized milk (IDF, 1972).

The joint FAO/WHO report on Milk Hygiene (1970) and the IDF Monograph (IDF, 1972) indicate that both biochemical and microbiological tests should be made to ensure that a satisfactory degree of sterility does exist in sterilized and UHT milk. The bacteriological test consists of incubating the milk sample in its original container and then subculturing on agar medium (0.1 ml of the incubated milk should give less than 10 colonies). Because the purpose of the test is to indicate the presence or absence of mesophilic bacteria able to grow in milk, a 2-class plan applies – a special situation not provided for in Table 11 for the cases associated with indicators.

The sampling plan recommended is a 2-class plan with $n = 5, c = 0$, the limits for the various tests being the same as those recommended by the International Dairy Federation (Table 23).

(c) *Ice Cream*

Ice cream is one of a class of products made essentially from milk and cream to which other ingredients such as sugar, stabilizers, gelatin, and flavouring are normally added ('simple' ice cream). Other specialty ingredients such as fruits, nuts, chocolate, meringue, biscuit, vegetable fats, synthetic sweeteners, etc. may also be added ('complex' ice cream).

Simple ice cream undergoes a heating process at a temperature at least equal to, but generally well above, that for HTST pasteurization of milk. The heating period is also generally longer. Present techniques of preparation, assuming minimal contamination after heat treatment and subsequent storage of the product in the frozen state, can provide 'simple' ice cream with very low numbers of viable bacteria, provided that the preparation is conducted in a closed processing cycle. However, inclusion of the specialty ingredients such as those mentioned above to make complex ice creams, or the external addition of decorations, meringues, biscuits, etc., will normally introduce other contaminants. Furthermore, the addition of such ingredients may cause a break in a closed system and provide other opportunities for the introduction of contaminants. Proposals for bacteriological limits must accommodate to these conditions. Accordingly, different sampling plans are recommended for 'simple' and 'complex' ice creams. Obviously, the manufacture of ice cream requires special care to select only ingredients of good bacteriological quality; this applies particularly to fruits and chocolate because they have been implicated in food-borne outbreaks.

At present, the bacteriological record of ice cream prepared with modern, industrial equipment is good (Borneff, 1963; Braga and Palladino, 1963; Cominazzini, 1964; Gillespie, 1963; Grosso, 1955), but the record of the small processing plants is less reliable. Consequently, only those firms which can obtain satisfactory and reproducible bacteriological results should enter international trade.

Ice cream is a product which may be included in either of the two main categories described at the beginning of this chapter. It may be on sale and consumed shortly after manufacture (i.e., in the category of products which must be controlled regularly on an a posteriori basis), but storage in the frozen state provides a shelf-life of several weeks or months. Hence, in the latter instance, analysis of the product provides a feasible method of quality control. Specific proposals are listed in Table 23.

The cases identified are: case 5 for the indicators and case 8 for staphylococci, with corresponding n and c values as suggested in Table 11, page 60. The M values for each test organism approximate upper limits of tolerance which are now used in several different countries. The limit for *Staphylococcus aureus* is well below that anticipated to cause enterotoxicosis, assuming consistent frozen storage. The m values represent levels known to be achieved by most manufacturers who consistently market ice cream of good quality. Occasional variation from such norms is accommodated by the c values specific for the plans.

Any involvement of ice cream in food-borne disease warrants immediate investigational sampling (see Section N, page 63).

(d) *Cheese*

Throughout the world diverse procedures are used to produce various types of cheese from milk which may be either raw, pasteurized (phosphatase-negative), or heat treated. The number of cheese types is large, and for many of them the available data concerning their bacteriological quality, from the hygienic point of view, are few and often have not been related to good manufacturing practice. More data are necessary before meaningful bacteriological limits can be recommended for every cheese variety. Nevertheless, for some of the cheeses most common in international trade, present knowledge and experience are sufficient to warrant proposals for bacteriological limits and sampling plans.

Although cheese has been shown to cause food-borne illness, epidemiological experience places cheese among the relatively infrequent vehicles of transmission.

1 *Aged cheese made from raw milk* Many varieties of cheese are made from milk which has not been heated before the cheese-making process begins. Certain types, however, require heat treatment of the curd at some point during the cheese-making process, e.g., during 'cooking.' Although many varieties of raw-milk cheese are made, the amount in international trade is relatively small.

Several pathogenic bacteria can survive for varying periods of time in cheese made from raw milk. Some measure of protection against the transmission of infectious disease, especially brucellosis and typhoid fever, is presently achieved by requiring cheese made from other than pasteurized milk to be aged prior to consumption for at least 60 days at a temperature not less than 4.4° c.

2 *Aged cheese made from milk heated during the cheese-making process* Some cheese types such as Gruyère, Emmental, and Grana are made from raw milk, but the curd undergoes a heat treatment of approximately 54–55° c for an hour or more during the cheese-making process. The application of these heating processes during cheese-making reduces the likelihood of bacteriological risk in cheese so made.

3 *Aged cheese made from pasteurized or 'heat treated' milk* Cheese is made also from pasteurized milk (phosphatase-negative) or milk heated to a degree which may not provide a negative phosphatase test, but does substantially reduce the number of microorganisms, including staphylococci and any other pathogens likely to be present in the milk. This latter process is sometimes referred to as 'heat treatment' or 'thermisation' of milk; the heat treatment approximates at least 65.5° c for 15 to 20 seconds.

Such heat treatments destroy or greatly reduce the number of heat-labile organisms, including pathogens. The presence of excessive numbers of such organisms in the final product is due either to contamination of the milk after heating, and subsequent growth, or to some faulty practice or accident during the cheese-making process, for example, starter failure (see below).

4 *Staphylococcal enterotoxin in cheese* The most serious health risk presented by cheese shipped in international trade is from staphylococcal enterotoxin, the result of growth of coagulase-positive staphylococci during cheese-making. The maximum growth of staphylococci occurs when whey drainage is complete (e.g., after pressing in the case of Cheddar and similar varieties). Later, during the weeks and months of the aging period, the staphylococci slowly die off and only a relatively small proportion of the earlier numbers will be recovered when cultured. However, any enterotoxins produced will remain.

Growth of enterotoxinogenic staphylococci to levels of at least one to 5 million per gram of cheese during cheese-making is necessary to produce enterotoxin detectable in cheese by the current assay method (Casman and Bennett, 1965). Such levels may result from faulty manufacturing practices. Heat treatments less severe than those indicated above, when applied to milk before or after the cheese-making process begins, may leave a residual population of coagulase-positive staphylococci which subsequently may increase to hazardous levels. In addition, insanitary equipment also may recontaminate the milk after heat treatment. However, the most important factor in the growth of staphylococci, and of many other undesirable bacteria, during cheese-making is failure of the lactic acid starter culture to develop normally ('starter failure'). Usually this occurs as a result either of infection of the starter culture by bacteriophage, or of the presence of small amounts of antibiotics or other inhibitory substances in the milk. Although bacteriological testing is advisable soon after manufacture, products in international trade usually cannot be sampled at the source of production. Therefore, any staphylococci found in ripened cheese probably represent only a fraction of the number present in the unripened cheese.

Available information about levels of staphylococci in fresh cheese is limited largely to hard and semi-hard varieties. With soft and semi-soft varieties, more information is necessary to determine (a) the disease hazard presented by such cheese contaminated with staphylococci and other undesirable organisms, (b) the levels of such organisms in the fresh product commensurate with good manufacturing practice, and (c) the changes in the number of staphylococci taking place during manufacture and ripening that may influence enterotoxin production. For the hard and semi-hard varieties, a viable count of 10,000 coagulase-positive staphylococci per gram of cheese appears to be well within technological achievement. Furthermore, this level in ripened cheese in international trade would seem to afford reasonable assurance that the number of coagulase-positive staphylococci did not reach a level associated with detectable enterotoxin.

Products likely to contain staphylococcal enterotoxin under conditions causing no change in concern would normally be allocated to case 8 (see Table 11, page 60).

Cheese suspected of containing enterotoxin should be tested preferably for the four more common enterotoxins, A, B, C, and D before release. Shortage of standardized type specific antisera preclude such tests for routine use in most laboratories.

5 *Recommendations* A 3-class sampling plan, $n = 10$, $c = 1$, is recommended; with the upper limit, M, of 10,000 viable *Staphylococcus aureus* per gram, and the lower limit, m, of 10^3/g. Further, an aging of at least 60 days at not less than 4.4° C (40° F) is recommended for all the hard and semi-hard cheese made from unpasteurized milk or curd and shipped in international trade.

Recently in the USA an outbreak of gastroenteritis occurred caused by soft cheese which contained pathogenic serotypes of *Escherichia coli*. This was the first recorded instance of such an outbreak attributed to cheese. While the coliform problem in cheese manufacture is well known, there are insufficient data available to support a coliform standard generally applicable to cheese. Accordingly, no sampling and testing plan is proposed at this time. Nevertheless, every effort should be made to adhere to manufacturing practices that will prevent coliform contamination and subsequent growth of these organisms during the manufacture of cheese.

C SAMPLING PROCEDURES

For general instructions on collecting and handling field samples, see Chapter 7. For specific procedures on sampling milk and milk products consult the International Standard (FIL-IDF 2:1958) proposed by the International Dairy Federation (IDF, 1958a). IDF specifications not included in the general procedures in Chapter 7 are outlined below, but it is recommended that the full IDF text be studied by all persons who collect samples of dairy foods.

(a) *Sterilized and* UHT *milk*

Always take original unopened containers as field samples.

(b) *Dried milk*

1 Where feasible, take original unopened packages as field samples.
2 When sampling bulk containers in the field or packages in the laboratory, use the precautions described in Section C(c), page 86 for opening the containers. Take the sample from a point near the centre of the container if possible. First remove the surface layer of powder with a sterile instrument (e.g., broad-bladed knife) and then draw the sample with a sterile spoon or probe (see Section B(a), 2, page 117).
3 In the event of dispute concerning the bacteriological condition of the surface powder in a package, take a sample of the product from this area.
4 Preferably use brown glass sample containers, to exclude light.

(c) *Ice cream*

1 Where feasible, take original unopened packages as field samples.
2 In sampling bulk ice cream, aseptically withdraw material from several points in the container, using a sterile spatula, spoon or trier. Transfer the ice cream to a sterile wide-mouth glass jar.
3 Keep field samples frozen until analysis has begun.

(d) *Cheese*

1 Use one of the following three techniques, depending upon the shape, weight, and type of the cheese: (*a*) sampling by cutting out a sector; (*b*) sampling by means of a trier; (*c*) taking a complete cheese as a sample. When a choice must be made between (*a*) and (*b*), method (*b*) is often more practicable, especially with hard cheese of large size.

(*a*) Sampling by cutting out a sector Using a knife with a pointed blade, make two cuts radiating from the centre of the cheese. The size of the sector thus obtained shall be such that after removal of any inedible surface layer, the remaining edible portion shall be more than twice as much as will be required for analysis. Use this method for Edam and Gouda cheese. It can also be used for semi-hard and soft cheese, and in general for all cheeses when sampling by means of a trier is impossible.
(*b*) Sampling by means of a trier The trier may be inserted obliquely towards the centre of the cheese once or several times into one of the flat surfaces at a point not less than 10 to 20 cm from the edge. From the plug

or plugs thus obtained cut off not less than 2 cm of the extremity containing the rind and use this piece to close the hole made in the cheese. The remainder of the plug or plugs shall constitute the sample.

Close the plug holes with great care, especially with large cheese, and if possible seal over with an approved compound. This method is most suitable for hard and semi-hard cheese, for example Emmental and Cheddar.

The trier may be inserted perpendicularly into one face and passed through the centre of the cheese to reach the opposite face. This method is suitable for Provolone, Caciocavallo, etc. In special cases a sector may be removed as in (*a*).

The trier may be inserted horizontally into the vertical face of the cheese, midway between the two plane faces, towards the centre of the cheese. This method is suitable for Tilsit, Cantal, Roquefort, Pecorino Romano, etc. In special cases a sector may be removed as in (*a*).

In the case of cheese transported in barrels, boxes, or other bulk containers, or which is formed into large compact blocks, sampling may be performed by passing the trier obliquely through the contents of the container from the top to the base. This method is suitable for processed cheese and cheese foods, etc.

(*c*) Sampling by taking an entire or substantial portion of cheese Use this method for fresh cheese (for example cottage, cream, or double cream cheese), for soft cheese of small size, and for wrapped portions of cheese packaged in small containers (e.g., some processed cheese and various soft cheese).

2 Immediately after sampling, place the samples (plugs, sectors, entire small cheese) in a sterile container of suitable size and shape, and seal. The sample may be cut into pieces for insertion into the container, but do not compress or grind it.

3 Send or transport the field samples immediately to the laboratory, and conduct the analysis as soon as possible, preferably on the same day. If either dispatch or analysis must be delayed, place the containers in a refrigerator at a temperature between 5° and 8° C. This is especially important with perishable cheese, e.g. soft cheese.

(e) *Precaution for analysis of sample units*

If bacteriological, chemical, and/or organoleptic analyses are to be made on the same field samples, always remove the sample units for the bacteriological tests first.

13

Sampling plans for raw meats

Six raw meat commodities, important in international commerce, were considered to be sufficiently involved in disease, particularly salmonellosis, to warrant microbiological analysis. The commodities were: poultry; comminuted meat; edible organs ('edible offal'); boneless meat, (i) for human consumption (pork, veal, beef, mutton), and (ii) for animal consumption ('pet meat,' including horse and kangaroo meat, fit for and sometimes eaten by man); and carcass meats, frozen and refrigerated. Boneless meat was divided into the classes (i) and (ii) because a higher proportion of pet meat samples have been found to contain salmonellae (Hobbs, 1965; Hobbs and Gilbert, 1970). Most of these meats are traded frozen.

The proposals for raw meats are based on data collected under a restricted range of circumstances, and the recommendations which follow are appropriate only to those data. The recommendations might consequently be inappropriate for different circumstances, and they are to be regarded for the present as suggestions only; more extended trials are required for definitive plans.

A SAMPLING PLANS

Table 24 lists the sampling plans proposed for immediate use, and those which would be desirable whenever the sanitary condition of the commodities in question improves sufficiently to invoke them without undue loss of food. They are restricted to tests for salmonellae and the standard plate count.

The use of indicator organisms such as *E. coli* was not recommended, because they occur so commonly in these foods, and also for reasons of economy. Nevertheless, the spread of enteropathogenic

TABLE 24

Sampling plans and recommended microbiological limits for raw meat and poultry

Product	Test	Method reference[a]	Case	Plan class	n	c[b]	Limit per g m	M
1 Carcass meat: chilled	SPC	64	1	3	5	3	10^6	10^7
	Salmonella	95	10	2	5	0	0	—
2 Carcass meat: frozen	SPC	64	1	3	5	3	5×10^5	10^7
	Salmonella	95	10	2	5	0	0	—
3 Boneless meat: frozen: (beef, veal, pork, mutton)	SPC	64	1	3	5	3	5×10^5	10^7
	Salmonella	95	10	2	5	1(0)	0	—
4 Boneless meat: frozen: (horse, kangaroo)	SPC[c]	64	1	3	—	—	—	—
	Salmonella	95	10	2	5	1(0)	0	—
5 Comminuted meat: frozen	SPC	64	1	3	5	3	10^6	10^7
	Salmonella	95	10	2	5	1(0)	0	—
6 Edible offal: frozen	SPC	64	1	3	5	3	5×10^5	10^7
	Salmonella	95	10	2	5	1(0)	0	—
7 Poultry: frozen	SPC[d]	64	1	3	5	3	5×10^5	10^7
	Salmonella	95	10	2	5	1(0)	0	—

a This column refers to page numbers in Thatcher and Clark (1968) where the test procedures are described. Use sample unit sizes recommended in the methods, except where otherwise indicated.
b Figures in parentheses in this column represent the desirable aim. Figures not in parentheses are an interim recommendation.
c SPC seems useful, but no data on SPC in these products have been found.
d Limits expressed for SPC refer to counts per ml of the original rinsing solution; plates incubated at 20° C.

types of *E. coli* from raw meat might be more important than hitherto understood (Shooter *et al.*, 1970).

Case 1 with a 3-class plan for the standard plate count at 35°C was chosen for all commodities, except horse and kangaroo meat, where the SPC data available were inadequate. For chilled (refrigerated) meats, plate counts at lower temperatures, e.g. 20°C, give a better prediction of the remaining shelf-life of the meat.

The *M* values for SPC represent an expression based on experience that meats with SPC values greater than 10^7 have been exposed to conditions favourable to spoilage; indeed, incipient spoilage may be detected. Further, most freshly imported meats show substantially lower values. The values for *m* represent knowledge of current commercial attainment based on large numbers of examinations of meats from different sources, and reflect experience that the tested meats with plate counts at or lower than *m* usually have not received abusive levels of contamination or unduly faulty handling. A normal shelf-life would be anticipated.

The present rate of contamination of raw meat and poultry with salmonellae is undesirably high (Hobbs and Gilbert, 1970; Hobbs, 1971; Crabb and Walker, 1970). Inadequate cooking of non-frozen material, or failure to increase the cooking time of frozen or partially thawed meat, may permit the survival of salmonellae (Pivnick *et al.*, 1968; Roberts, 1972; Todd and Pivnick, 1973). Multiplication of these pathogens in the cooked food could occur at ambient or warmer temperatures. Even though meat is usually cooked enough to destroy salmonellae, the possibility still remains that these organisms from raw meat may be spread to cooked products by the hands of food handlers, or by cutting utensils and other equipment and surfaces (Gilbert, 1970; Gilbert and Watson, 1971). In addition, food handlers may become infected, and those with diarrhoea may excrete large numbers of salmonellae. Efforts to bring about control must in the end be based on livestock handling, but, for immediate action, condemnation of the worst supplies of raw meats and poultry should lead to prompt improvement.

The presence of salmonellae in these raw foods is a disease risk. Cooking should reduce the hazard and therefore case 10 is applicable; nevertheless cross-contamination might occur after cooking.

It would be desirable to have all raw meats, poultry, and edible organs – whether frozen or not – free from salmonellae so that $c = 0$ would apply; yet because even $n = 10, c = 1$ would reject a commercially intolerable number of samples, at present the Commission is forced to accept $n = 5$, $c = 1$. For the future we contemplate more stringent sampling plans, such as $n = 5, c = 0$, and later $n = 10$ (see Table 24).

B SAMPLING PROCEDURES

(a) *Wholesale cuts or packages*

The sampling procedure for chilled or frozen carcass meat and boneless meat in bulk entails the examination of 5 field samples (Table 24) (see page 83 for distinction between field sample and sample unit). Take one field sample from each of 5 separate carcasses or packages of meat, and if the lot is distributed in several shiploads or freight cars, take the field samples from more than one transportation unit.

Except for poultry, it is impractical to examine whole carcasses and therefore field samples should be taken from the most contaminated areas. From carcasses of cattle and sheep, take field samples from the sticking, prepectoral region, thin flank, sacral region, goose skirt, renal area, and neck; from pig carcasses take field samples from the neck and behind the ears (Hobbs and Wilson, 1959). Preferably the field sample should consist of equal amounts of meat (about 200 g) from each of these areas. Where defacing of the carcass is economically undesirable, other methods of collecting samples, such as the surface swab technique, the spray-gun method (Clark 1965a, 1965b) or rinsing of the entire carcass may be used. In all cases the various portions from each carcass are pooled and thoroughly mixed in the laboratory to form a composite field sample from which one sample unit is taken for the Standard Plate Count and one sample unit for each of two *Salmonella* tests. In the Standard Plate Count analysis, counts for replicate plates are averaged to give one result for each carcass.

In collecting field samples, remove wrappings from the carcass or package carefully without handling the meat. When removing portions of meat, use sterile instruments (knives for unfrozen meat; saws and special drills for frozen meat) and transfer the piece aseptically to sample containers. To avoid cross-contamination use separate instruments for each carcass or package of meat; or clean and sterilize the instrument between samples (see Section c(b), page 85 for instruction on sterilizing instruments in the field). Place portions of meat from each carcass or package into a single sample container, preferably a double plastic bag (one inside the other), seal, label, and transport to the laboratory as described in Section c(c), page 86. Samples of frozen meat should remain frozen while samples from chilled carcasses should not be frozen.

(b) *Retail packages*

The sampling procedure for fresh or frozen raw meats in retail packages or for fresh or frozen poultry entails the examination of 5 packages or 5 poultry carcasses, each from separate containers in the lot. When sampling poultry, the 5 packages can be taken from the same carton because of a presumed heterogeneity among the individual birds in a lot.

Mark each package with the required information as described in Section C(e), page 85 and transport it chilled or frozen to the laboratory as described in Section C(g), page 88.

C TEST PROCEDURES

For analysis, comminute and blend the various portions of meat composing the field sample. From this composite field sample weigh out sample units for the SPC and *Salmonella* tests according to directions given in Thatcher and Clark (1968). If the field samples are frozen packages, thaw them overnight in a refrigerator at 0–5°C before preparing the composite sample. The 'drip' obtained on thawing can be used for the Standard Plate Count for frozen boneless meat; use 1 ml of drip for each sample unit.

To obtain a test specimen for determination of SPC and for isolation of salmonellae from poultry, a modified technique from Surkiewicz *et al.* (1969), is recommended: shake a carcass or portions thereof in a plastic bag containing 300 ml of lactose broth; use 1 ml of the broth for determining the SPC; and add the remainder to 300 ml of double strength liquid enrichment medium for *Salmonella*.

For foods other than poultry, the test for *Salmonella* requires a sample unit of 25 g and is based on positive or negative results using Procedure III (Thatcher and Clark 1968, page 95). Conduct the test in duplicate for each field sample. For frozen products, use a non-selective enrichment: incubate 25 g sample units for 16 to 24 hr at 36°C ± 1°C in 100 ml of lactose broth; then transfer 10 ml of the resultant culture to 100 ml of the selective enrichment broth used and incubate at 43°C for 24 hours before subculturing on the agar media (Edel and Kampelmacher, 1972).

14

Sampling plans for processed meats

According to the FAO Trade Yearbook (1971), international trade during 1970 of meat in airtight containers and meat preparations in, or not in, airtight containers amounted to about 800,000 metric tons, while the volume of sausages in international trade was about 70,000 metric tons.

The processed meats currently in international trade are largely restricted to four groups of products: (a) shelf-stable cured or uncured canned meats; (b) perishable cured canned meats; (c) cooked, uncooked fermented, or dried or semi-dried sausages; (d) sliced cured meats packed under vacuum.

The discussion of shelf-stable canned foods in general (Chapter 15) is pertinent to the canned foods considered below. To assure the safety of such foods, the advantage of in-plant control over terminal analysis is again emphasized, as in Chapter 15.

Microbiological examination of the final product should be undertaken only for non-shelf-stable (perishable) cured meats and only if inadequate processing or temperature abuse during transportation is suspected. The test organisms would be the indicators, *Bacillus* species and non-sporing mesophiles. The latter, with the possible exception of *Micrococcus* and enterococci, when present in perishable canned meats indicate an inadequate heat process or post-heating contamination. However, the heat process is not the only safety factor to be considered. For a discussion of the microbiology of perishable canned meat products, see Report of the First International Symposium of Food Microbiology (1955). See also Section A below.

A SHELF-STABLE CURED OR UNCURED MEATS

Cured, canned meats (comminuted as in luncheon meat and canned sausages, and meat in whole pieces like hams in consumer-size cans) move frequently in international trade. Because canned, uncured meat products also play a rôle in international trade, they are considered together with the cured meats in sampling plans.

To control the safety and stability of shelf-stable canned meat products solely by microbiological tests on the end product would require examination of a tremendous number of cans – far more than existing laboratory facilities and personnel could handle on a routine basis. Moreover, better knowledge of safety and stability could be gained if data on production control were put at the disposal of the controlling agency.

It is therefore recommended that data from in-plant control should be made available to the controlling agency, as emphasized in the discussion of sampling plans for canned foods. Safe processing of cured canned meats is based on a complex relationship between content of water, salt, and nitrite, pH, number of contained spores, and process value (a temperature-time function). All such data should be within the limits of known good commercial practice before routine acceptance of a lot. If such data are adequate and satisfactory, further testing of the finished product is considered unnecessary.

If these data, however, are not available or are inadequate, sampling should take place, according to the sampling plan for canned foods, i.e., visual inspection for swells and seam defects. Under certain circumstances cans are also examined after incubation at 30–37°C.

If swells are noted among the cans investigated, specimens should be sent to a laboratory to evaluate the probable cause of the defect. If the examination reveals spore-forming organisms, under-processing is likely to be the cause; the presence of non-sporing organisms (with the possible exception of *Micrococcus* species) with or without a seam defect is indicative of leakage.

Within the laboratory, exacting control of laboratory technique, cleanliness of the facilities and especially of the bacterial quality of incoming air are important in order to minimize 'false-positive' tests, i.e., the reporting of the presence of mesophilic spores which may have been contributed from within the laboratory.

B PERISHABLE CURED CANNED MEATS

The most important meat products of this kind in international trade are pasteurized canned cured hams and pork shoulders. While the importance of testing such products is recognized, nevertheless the high price of canned hams and similar foods is a serious obstacle to intensive examination. For these products, data on thermal processing, water

TABLE 25

Sampling plans for perishable canned cured meats (to be used when processing data are not available or are unsatisfactory)

Tests		Method reference	Case	Plan class	n	c	Limits: count per g		Acceptance or rejection criteria
							m	M	
Step 1	Visual inspection for swells and seam defects		—	2	10	0	—	—	Reject if 1 or more defective cans are found. Proceed to step 2, if no defective cans are found
Step 2	Measurements of air temperature between cans		—	2	10	0	—	—	Accept if temperature is below 10° C. If higher proceed to step 3
Step 3	Determination of SPC in meat and jelly	ICMSF[a]	2	3	5	2	10^3	10^4	*m* and *M* values refer to *Bacillus* spp. For other non-sporing organisms do not accept. See Section B, this chapter, for interpretation of the presence of *Micrococcus* and enterococci

a Thatcher and Clark, 1968, p. 66. Use sample unit size recommended in the method.

supply, seam inspection, and chemical composition should be available, as for shelf-stable canned meats (page 142), together with records of temperature during shipping. If all such data are satisfactory, no testing is necessary.

(a) *Sampling procedure in port of entry*

If adequate data are not available, sampling should be done according to the following scheme (see also Table 25).

1 The sample consists of a total of 10 cans that should be drawn at random from at least 5 different cartons or containers. Examine these visually for 'swells' and seam defects. If no defective cans are found, pass the lot. If one or more defective cans are found, reject the lot.

2 The temperature on the surface of the cans should be measured while they are being drawn from the carton, preferably with an electronic temperature measuring device with a suitable measuring unit. This temperature should not be higher than 10°C and ideally should be lower. If the temperature is below 10°C, pass the lot. If not, record the temperature.

3 If the temperature anywhere in the lot is higher than 10°C, take 5 cans from the warmer places for a microbiological examination, identify them as described in Section C(e), page 88, and cool and transport the cans to the laboratory as described in Section C(g), page 88.

(b) *Laboratory analysis*

At the laboratory, sample units are drawn with aseptic precautions from each can, following the procedure for sampling the can contents as described by Hersom and Hulland (1969) in order to obtain one sample unit from the centre and one sample unit from the jelly of each can. Examine these 2×5 sample units for aerobic plate count. Use sampling plan case 3, $n = 5, c = 2$ for $m = 10^3/g$, $M = 10^4/g$.

Use the higher figure from the two specimens of each can to determine whether or not there is a defect. The m and M values refer to *Bacillus*. The presence, in any sample, of non-sporing organisms other than *Micrococcus* or enterococci indicates a defect. Some strains of *Micrococcus* and enterococci have been reported capable of surviving the heat treatment normally given these foods. Accordingly, these organisms may be tolerated in small numbers.

It should be emphasized that the values for m and M in this context do not necessarily reflect any spoilage or health hazard, but merely indicate degrees of undesirable manufacturing practice. The limits have been established from data derived from very large numbers of analyses over several years and are known to be obtainable under acceptable practice. If the counts indicate a defect in the lot, the decision on the action to be taken should be left to the discretion of the authority concerned.

C SAUSAGES

The several hundred varieties of sausages could be divided into three groups: (i) cooked, (ii) uncooked fermented, and (iii) dried or semi-dried. However, many of these types are specialities produced and sold locally. They move to a very limited extent into international trade.

Because of lack of knowledge of the products found in international trade, and because of the diversity in manufacturing practice and composition (e.g., heat treatment, ultimate acidity, and water activity), microbiological criteria for these products will not be proposed, though at this time consideration regarding sampling plans for pathogenic organisms is probably desirable.

D SLICED CURED MEAT PACKED UNDER VACUUM

Some of these products have been shown to allow production of botulinus toxin (Pivnick and Barnett, 1965) and growth of salmonellae (Davidson and Webb, 1972). Experience in some countries indicates that they are often abused through inadequate refrigeration. Many provide a suitable substrate for the growth of several food-poisoning organisms; nevertheless, they seldom are implicated in disease.

The Commission does not know of any substantial international trade in these products (e.g., vacuum-packed sliced bacon, sliced ham, etc.). If the trade should increase, however, it would be advisable to study the microbiology of these products more closely.

It is assumed that sliced bacon will be heated by the consumer and, in addition, the growth of lactic acid bacteria is expected to afford protection against the growth of *Staphylococcus aureus*. To date, there is no evidence of food-poisoning from sliced bacon marketed under normal conditions. Accordingly, the need for protective analysis is not considered urgent.

Sliced ham and similar products will in most cases be consumed without further treatment. These products should be considered under case 3 (see Table 11, page 60) for the aerobic plate count, and under case 8 for *Staphylococcus aureus* because that species will not multiply under refrigerated storage and does not readily produce toxins in the ecological conditions found in vacuum-packed cured meats.

15

Sampling plans for shelf-stable canned foods

Shelf-stable canned foods are those which (i) have been packed in hermetically sealed containers and (ii) have been through a heat process adequate either to destroy all of the microorganisms present or to ensure that any surviving organisms do not grow in the product. Such processes include: (a) the so-called 'botulinum cook' treatment appropriate to low-acid foods (i.e., pH 4.5 or above); (b) the lesser heat treatment applied to products containing certain curing ingredients; and (c) the mild heat treatment given to products of pH below 4.5 ('high acid' foods). Such products have been termed 'commercially sterile,' but are more properly designated 'correctly appertized' (Goresline *et al.*, 1964).

Canned foods could have the potential to cause the often fatal disease, botulism, but experience demonstrates that, if present, *C. botulinum* would be expected to occur at such a low frequency that no conceivable sampling plan would be adequate as a direct measure of its presence.

Accordingly, indirect control is applied, consisting of (i) 'in-plant' control to assure the use of an effective process, (ii) good plant sanitation, and (iii) limited terminal analysis to establish whether the process has indeed been effective and whether the can seams are capable of maintaining the 'hermetically sealed' condition. Introduction of spoilage and pathogenic organisms from polluted water through seams has occurred occasionally. Frequent checking of the cooling water and of seam structure would minimize this risk.

A A POLICY FOR COMMERCIAL PROCESSING OF CANNED FOODS

As a result of the two widely publicized botulism episodes concerning canned soups in the United States in 1971, the National Canners Association filed with the United States Department of Health Education and Welfare (USFDA, 1971) a proposed statement of policy on the commercial processing of foods in hermetically sealed containers (see Appendix 5). A similar code of practice should be considered for all canned low-acid vegetable products entering international commerce. The adoption of such a code would maintain and enhance confidence in the safety of all such products.

B INTEGRITY OF CAN SEAMS

Seam integrity of hermetically sealed containers used for canned foods is critical to safe processing and requires constant surveillance by the can manufacturer and canner. A defective can may permit recontamination of the can contents following the heat process. If the contaminant is a pathogen, a health hazard exists. Experience has shown, however, that such hazards are rather remote, notwithstanding recorded illnesses as a result of defective seams (type E botulism from canned tuna (Johnston *et al*, 1963); typhoid from canned corned beef (Anderson and Hobbs, 1973); staphylococcus food-poisoning from canned peas (Bashford *et al*, 1960)).

To minimize the chances of spoilage caused by defective seams, the can manufacturers, as well as the canners, should maintain a strict quality control program of inspection and of can seam measurements to ensure that predetermined tolerances are met. A statistical approach employing an attributes sampling plan is usually followed.

C COOLING WATER

In addition to the can quality control program, the canner should use only chlorinated potable water for cooling the retorted cans. A sampling plan for examination of the cooling water should be the responsibility of the manufacturer.

D INCUBATION TESTS

In some instances, results of incubation tests of a small proportion of cans from each lot should be made available to the controlling agency. However, these tests are of little or no value in assessing safety in those foods that have received the so-called 'botulinum cook,' where reliance should be placed upon determining the adequacy of the thermal process from the data provided. On the other hand, incubation tests are desirable for some low-acid canned foods, such as cured meats, which do not normally receive a calculated 'botulinum cook,' but are otherwise known to be relatively stable because they not only undergo a heat process but have curing agents added as well. For such canned foods, processed in a so-called still retort, incubation tests should be performed on at least two cans obtained from two different positions in each retort, for example, one from the centre, the other from the top.

It is recommended that a small portion of all canned low-acid foods that have been heat processed by one of the newer high temperature–short time (HTST) methods (continuous or discontinuous agitating retort, aseptic process, etc.) be subjected to an incubation procedure by the processor and the results made available to the controlling agency. As a minimum program, it is recommended that at least one can be obtained from each line in operation every 15 minutes, be identified as to time of removal, and be incubated for 14 days at 30–37°C. Following incubation, all cans are inspected for swells, then approximately 10 per cent should be opened for thorough inspection of the container and contents. If no swells occur, and if the contents of all opened cans show no evidence of spoilage or pH change the lot may be considered satisfactory. Again, however, it should be emphasized that no sampling procedure assures adequate safety with respect to the botulism hazard – greater reliance should be placed upon thermal processing records, constancy of formula, and the use of instrumented fail-safe procedures to signal any interruption of the proved process parameters.

E PROCESS CONTROL DATA: AVAILABILITY TO CONTROL AGENCIES

In addition to data on sampling, incubation, and thermal processing, it is recommended that information on water supply, seam inspection, appropriate chemical composition, and pH be made available to the controlling agency. The data on chemical composition should give, at least, mean values and standard deviation or range. If such data are adequate and satisfactory, further testing of the finished product is considered unnecessary. However, if these data are not available or are inadequate, samples should be taken. All should be taken at random (see Section B, page 10).

F IMPORT INSPECTION

Microbiological examination of canned foods is not recommended on a routine basis. Instead, visual inspection of the cans at the port of entry

TABLE 26

Sampling plans for shelf-stable canned foods (to be used when processing data are not available or are unsatisfactory)

Tests		Method	Plan class	n	c	Limit: number of defectives		Acceptance or rejection criteria
						m	M	
Step 1	Visual inspection for swells and seam defects		2	200	—	0	—	Accept when no defective cans are found; if 1 or 2 defective cans are found proceed to step 2; if 3 or more defective cans are found, reject
Step 2	Sorting of the whole lot when $c = 1$ or 2		2		1%	0	—	If less than 1% show visible defects proceed to step 3. Reject if 1% or more of the cans are defective
Step 3	If less than 1% show visible defects, test 200 sound cans	Incubate at 30–37° C for at least 10 days. Observe swells	2	200	0	0	—	Reject if 1 or more swells are found after incubation: if no swells are found proceed to step 4
Step 4	Examine 20 random incubated cans for seam defects	Tear seams. Inspect as suggested by can manufacturer, or USFDA (1971)	2	20	0	0	—	Accept if no defective seams are present in incubated sample

should be carried out, and under certain circumstances (see below) incubated cans should be observed for swells and seam defects.

The initial inspection of a lot for swells and seam defects should be carried out on 200 cans (see Table 26) taken at random from cartons chosen also at random throughout the lot. Cans are selected randomly from the shipping containers in accordance with the following schedule.

No. of cans per carton	*No. of cans to pick*
5 or less	all
6 – 12	6
13 – 60	12
61 – 250	16
251 or more	24

If, for example, each carton contains 24 cans, open 17 cartons at random, and pick 12 cans at random from 16 of them, and 8 cans at random from the remaining carton (total, 200 cans). Draw more cans from each carton if the number of shipping containers in the lot is not sufficient to yield 200 cans in accordance with the above schedule. Select all cans in the lot if the total number of cans is less than 200. Identification of individual cans at this point is unnecessary.

Examine the 200 cans for 'swells' and seam defects. For the purpose of this examination, 'swells' refers to cans which have become distended by gases presumably brought about as a result of microbial action within the product; they are not cans with dents. If no defective cans are found, the lot is accepted, and if three or more are found, it is rejected.

If one or two defective cans are detected, the whole lot may be sorted for the removal of defective cans, and the sampler or a person authorized by him must be present during the sorting to remove, count, and identify the defective cans. If this sorting reveals 1% or more of visibly defective cans, the lot is rejected. If less than 1% are defective, 200 of the sorted, sound cans are taken at random and incubated, according to the aforementioned schedule.

If the 200 cans chosen for incubation are placed in shipping cartons, no identification of the individual cans is necessary, but a label should be affixed to the carton, as described in Section C(e), page 88.

Refrigeration is not necessary during transportation of the cans to the laboratory. Upon arrival the cans should be removed from the carton and each can properly identified.

For low-acid foods and cured meats, incubate the cans at 30–37°C for at least 10 days. If thermophilic spoilage is of concern, incubate one half of the cans for the same period at 55°C ± 1°C. For acid canned foods, use an incubation temperature of 25°C ± 0.5°C. Under some circumstances, longer incubation periods may be appropriate for all these classes of foods. A daily check of the cans for 'swelling' is advisable.

If any of the cans incubated show swells, reject the lot. If no swells occur, choose 20 cans at random and examine them for seam defects by tearing the cans down. Inspect the seams in accordance with procedures described by USFDA (1971) or by can manufacturers. For some foods, pH measurements on the contents would be appropriate. If none show defects, accept the lot. If among the cans investigated swells appear, send those cans to a laboratory for evaluation of the probable cause of the defect.

The acceptable quality level (AQL) for $n = 200$, $c = 0$ is 0.025%; that is, lots with 0.025% defective cans will be accepted 95% of the time. Lots with 0.1% defectives will be accepted about 82% of the time, lots with 0.35% defectives will be accepted about 50% of the time, and lots with 1.15% defectives will be accepted only 10% of the time (consumer's risk). The plan $n = 200$, $c = 0$ should be considered minimal in stringency. Some manufacturers express concern at a finding of 0.01% defectives. It is again emphasized that the proposed sampling scheme is not intended, and cannot suffice, as a safeguard against the presence of *C. botulinum*. It would, however, detect an important fault in processing or seam closure.

16

Microbiological evaluation of fresh or frozen raw shellfish

Molluscan shellfish, such as oysters, clams, and mussels are sedentary aquatic animals that feed by pumping relatively large volumes of water through their gills. In the process, they accumulate microbiological and chemical pollutants from the water, which are, in turn, transmitted directly to the consumers of raw whole shellfish. Public health control of these products depends primarily on sanitary survey and patrol of shellfish-growing areas and on inspection of harvesting, processing, and distribution operations. Microbiological examination of shellfish serves mainly to confirm the effectiveness of these sanitary practices by the industry, and it has little meaning unless the source and sanitary history of the lot are known.

When laboratory tests are used to check periodically on shipments from growing areas approved by a regulatory authority, relatively small numbers of sample units per lot can yield useful information. Many control laboratories in the US and Canada may examine only one or two sample units per month from each shipper. This minimal laboratory effort has been used successfully for nearly fifty years in conjunction with sanitary surveys and inspections to monitor the safety of shellfish harvested from approved growing areas by certified shippers; however, it is not applicable to shellfish of unknown origin.

A SAMPLING PLANS

Shellfish are recognized as a special class of food for several reasons: (i) their growth habitat is often polluted coastal waters; (ii) the areas of non-polluted coastal waters are rapidly diminishing; (iii) shellfish from polluted areas have been implicated in several outbreaks of typhoid fever, other salmonelloses, and infectious hepatitis; (iv) various methods of self-cleansing (depuration) of shellfish are used in commercial practice; and (v) the control of raw, fresh, or frozen shellfish in international commerce involves a complex system embodying (a) sanitary survey and patrol of growing areas, (b) plant inspection in accord with defined codes, and (c) bacteriological limits for the marketed products.

As an illustration of this system, reference is made to the bilateral shellfish agreement between Canada and the United States which has served as a guide for agreements with other nations such as Japan and the Republic of Korea. The applicable sanitary and testing specifications are described in the National Shellfish Sanitation Program Manual of Operations (Houser, 1965). Two portions of Part I of the Manual: Section C, 'Growing area survey and classification' and Appendix A, 'Bacteriological criteria for shucked oysters at the wholesale market level' are included here as Appendix 6. This plan is, of course, different from the three-level sampling plan referred to elsewhere in this text.

The microbiological criteria for approved growing areas are based on a coliform median MPN (most probable number) of the overlying water that does not exceed 70 per 100 ml. No more than ten per cent of the water samples may exceed a coliform MPN of 230 per ml when determined by a five-tube decimal dilution test.

Wholesale market standards for fresh or frozen oysters from approved growing areas are based on a faecal coliform content of no more than 230 MPN per 100 grams of homogenized shell liquor and meats, and on a 35° C plate count of no more than 500,000 bacteria per gram. Oysters that do not exceed these limits are considered 'satisfactory.' Those above the limit are regarded as 'conditional' and may be accepted or rejected by regulatory authorities. If two or more successive samples from the same shipper exceed either standard, further shipments may be prohibited by the regulatory authority until appropriate corrective actions have been taken by the shipper.

Somewhat similar systems for sanitary control are employed in certain other countries, but they utilize different criteria and methods than the North American system. Wood (1970) has described these differences with respect to control of British and French shellfish. Undoubtedly, each system is useful in the area of its intended application, but agreement on a single system for worldwide adoption is not possible at this time.

Uniform sampling plans for the microbiological analysis of shellfish would be desirable. None are proposed, however, because of the lack of consensus and the complexity of the factors, other than microbiological, that determine acceptance or rejection of international shipments. Sampling plans are also needed for examination of shellfish from unknown growing areas which cannot be banned from the marketplace in some countries unless they are shown to be potentially hazardous to health. Unfortunately, even less agreement exists on the criteria for this purpose.

The importance of sanitary surveillance and control of shellfish-growing areas, depuration operations, and processing plants cannot be overstated. The Commission recommends strongly that all molluscan shellfish moving through international trade channels be harvested only from sources under continuous surveillance and control by a reputable agency or organization.

B SAMPLING PROCEDURES

When samples of shell stock and shucked unfrozen shellfish are collected for microbiological examination, they should be held under dry refrigeration at 1 to 10° C and examined within approximately six hours after collection. Frozen shellfish may be examined after more protracted periods of storage, provided they are kept continuously frozen. Each sample unit should include a sufficient number of shellfish to yield approximately 200 grams of homogenized shell liquor and meats. Details of the procedures used in the US and Canada for collection and analysis of samples are described in Recommended Procedures for the Examination of Sea Water and Shellfish (APHA, 1970). British and French procedures have been noted by Wood (1970).

Conclusions

The recommendations of this book provide, for the first time, an opportunity to apply a system of international surveillance of foods through the use of standard sampling plans. They also provide a rational basis for microbiological standards, which should be made more or less stringent as circumstances warrant. The specific recommendations express the principle that stringency of examination of a food should be related to the severity of the hazard to be anticipated under normal conditions of use.

The proposed criteria of acceptance are within the current capabilities of competent food processing and, if generally met, would do much to secure food safety and remove spoiled products from international trade. However, the microbiological limits specifically recommended are all subject to change based on new survey data and changes in technology. In the meantime, the proposed plans represent the best consensus of judgment in the light of the total data and experience available to the Commission (see pages 81–2).

The plans go beyond most contemporary practices by specifying for each the number of sample units to be examined and rational acceptance criteria, besides a procedure for obtaining samples, thus allowing a definite statement of the consumer's and the producer's risks associated with each plan.

The adoption of 3-class plans is an important innovation which, by accommodating a tolerable number of microbiological values that may appear to exceed an 'ideal' limit, recognizes normal distribution ranges and thereby removes a major prejudice against the use of microbiological criteria for regulatory purposes.

The stringency of most of the plans is limited to an undesirable extent by the number of specimens which it is practicable to examine in laboratories, and in some instances by the ability of producers to meet the more stringent criteria of preferable plans. Reduction of the consumer's risk may well be feasible in the near future if extended experience warrants further use of composited sample units, as proposed for *Salmonella* tests (see page 70). Such aggregation permits analysis of larger numbers of sample units at relatively little increase in cost, thereby removing a major impediment to increased plan stringency.

The proposed plans have the advantage of detecting seriously substandard lots at much higher probabilities than lots of marginal quality. Nevertheless, they grant, on a short-term basis, only partial consumer protection. Control of processing and handling at the source offers better consumer protection. The long-range prospects for protection by sampling at ports of entry, however, are substantial. Even if the probabilities associated with a particular plan cause rejection of only 1 in 20 of lots marginally substandard to the criteria of acceptance, this, nevertheless, represents a severe penalty to the processor, who would normally make an effort to improve, as needed, his processing, sanitation, or control methods. Indeed, at present, the primary need in the attainment of food safety is improved plant quality-control in all its facets – environment, refinement of process, sanitation, and control-testing. In the meantime, microbiological analysis of the product on receipt remains essential and the tendency to increase it will continue.

Since economic factors impose the major restriction to analysis, there is an urgent need for research to develop faster analytical methods, using the potential advances provided by modern laboratory technology and instrumentation. Most current methods for enumerating microorganisms are largely refinements of the earliest isolation procedures. The aim should be to create procedures to enable analytical centres to examine very large numbers of samples rapidly and at low cost. Some of the laboratories with which Commission members are associated have already begun such development.

The objectives of this book as specified in Part I are substantially met. No longer should shipments of food be accepted or rejected on the basis of casual and unsatisfactory sampling. Indeed, several of the agencies and companies represented by Commission members, and food manufacturing companies whose financial support gave them access to earlier drafts of this book, have already begun to implement the Commission's proposals. The 3-class plans have gained wide support. Nevertheless, lack of adequate data made it unwise to propose sampling plans for all foods important in international trade. To overcome this deficiency, the Commission has initiated a long-term computer-programmed study to collect, from available sources throughout the world, data on the microbiological content of foods, and to estimate their relation to such factors as disease involvement, quality of practice in production and processing, and the effects of control analysis on food quality and health. The purpose is to establish a more secure basis for choosing microbiological criteria and for devising additional or amended sampling plans which may be required in the future.

Other acceptance procedures will doubtless be examined, as data accumulate to show that a food manufacturer or class of manufacturers can consistently meet acceptable microbiological criteria. A probable alternative might be the use of variables plans applied periodically to justify a normal official acceptance of lots based on the manufacturer's own control data.

In the meantime, even a sound and realistic set of guiding principles has little value unless it is supported by sincere effort towards practical fulfilment by manufacturers and rational enforcement by the agencies responsible. Administrators of regulatory agencies will not protect consumers if they fail to accept laboratory findings and – whether by design, or by ignorance – fail to act on them.

APPENDICES

Appendix 1

PROGRAM OF THE INTERNATIONAL COMMISSION ON MICROBIOLOGICAL SPECIFICATIONS FOR FOODS

The overall purpose is to appraise public health aspects of the microbiological content of foods, particularly those of international interest, and to make appropriate recommendations to aid in establishing internationally analytical methods and guides to interpretation of the significance of microbiological data. It is hoped that developing countries will thereby derive particular benefit even though the initial motives for organization of the Commission were based upon existing needs within countries with more advanced food technology.

The Commission has agreed upon the following specific functions:

1 To recommend microbiological criteria of acceptance for pertinent categories of microorganisms in specific foods.
2 To recommend defined methods of analysis and sampling, for detecting and enumerating food-borne microorganisms for which criteria of acceptance are proposed
3 To establish an international system of collaborative testing of methods, for the recommended microbiological examination of foods, with special application to criteria of acceptance or to microbial species of immediate public health concern
4 To establish channels of organization to aid international exchange of information on methods in microbiological analysis of foods
5 To establish a computerized data bank on food microbiology appropriately programmed to aid in establishing rational criteria of acceptance and other correlations pertinent to food-borne disease and utility: the data to be derived wherever available throughout the world and, in due course, to be available on request to groups having appropriate professional interest
6 To consider the desirability and feasibility of establishing an international centre or centres for reporting outbreaks of food-borne illnesses from *Salmonella, Vibrio comma,* shigellae, *Clostridium botulinum*, and enteropathogenic viruses, and others as they may become of critical concern
7 To discuss and formulate 'Group Statements' on the significance of specific pathogenic bacteria in foods, e.g. salmonellae in animal feeds and feed ingredients; staphylococci in dairy products; type E, *Clostridium botulinum* in fish and fish products; *Vibrio parachaemolyticus* in fish and shellfish; viruses in foods

8 To serve as a consultative body of Food Microbiologists to offer advice, on request, to International Agencies with regard to public health aspects of the microbiological content of foods
9 To forward the recommendations of the Commission, through its parent body, IAMS, to pertinent International Agencies such as WHO, FAO, UNICEF, as expedient
10 To maintain liaison with other allied committees with a common interest such as the committees now engaged in development of the Codex Alimentarius, AOAC, the Protein Advisory Group of UNICEF
11 To recommend areas of research for the solution of specific problems in food microbiology, e.g., (i) the effect of selective antimicrobial agents, such as antibiotics, on microbial populations in foods; (ii) the effect of new processing or packaging systems on the microbial ecology of foods; (iii) the effect of non-lethal irradiation treatments (radurisation, radicidation) on selection and mutagenesis in food-borne microorganisms; (iv) the effect of destruction of the normal flora by heat or other agents on subsequent development of pathogenic microorganisms; (v) the effect of irrigation practices on contamination of raw foods
12 To establish and maintain liaison between the Commission and sources of Research Grants for microbiological studies within the scope of the Commission

Appendix 2

MEMBERS, SUBCOMMISSION MEMBERS, AND CONSULTANTS OF ICMSF

MEMBERS (current, 1977)

Dr H. LUNDBECK (*Chairman*), Director, The National Bacteriological Laboratory, S-105 21, Stockholm, Sweden

Dr D.S. CLARK (*Secretary-Treasurer*), Chief, Research Division, Bureau of Microbial Hazards, Food Directorate, Health Protection Branch, Health and Welfare Canada, Tunney's Pasture, Ottawa, Ontario, Canada, K1A 0L2

Dr A.C. BAIRD-PARKER, Head, Microbiology Research, Unilever Research, Colworth/Welwyn Laboratory, Unilever Limited, Colworth House, Sharnbrook, Bedford, England, MK44 1LQ

Dr FRANK L. BRYAN, Chief, Foodborne Disease Training, Instructional Services Division, Bureau of Training, Center for Disease Control, Public Health Service, Department of Health, Education and Welfare, Atlanta, Georgia 30333, USA

Dr J.H.B. CHRISTIAN, Associate Chief, Division of Food Research, CSIRO, P.O. Box 52, North Ryde, NSW 2113, Australia

Professor C. COMINAZZINI, Director, Department of Microbiology, Provincial Laboratory of Hygiene and Prophylaxis, Viale Roma 7d, 28100 Novara, Italy

Professor OTTO EMBERGER, Chief, Department of Microbiology and Associate Professor of Hygiene, Faculty of Medical Hygiene, Charles' University of Prague, Srobárova 48, Praha 10, Vinohrady, Czechoslovakia

Dr BETTY C. HOBBS, Microbiology Department, Christian Medical College and Brown Memorial Hospital, Ludhiana, Punjab, India

Dr KEITH H. LEWIS, Professor of Environmental Health, School of Public Health, University of Texas, Health Sciences Center at Houston, P.O. Box 20186, Houston, Texas 77025, USA

Dr G. MOCQUOT, Chargé de mission à l'INRA, Technologie laitière, CNRZ, 78.350 Jouy-en-Josas, France

Dr N.P. NEFEDJEVA, Chief, Laboratory of Food Microbiology, Institute of Nutrition, AMS USSR, Ustinsky pr. 2/14, Moscow G-240, USSR

Dr C.F. NIVEN, JR., Director of Research, Del Monte Research Center, 205 North Wiget Lane, Walnut Creek, California 94598, USA

Dr P.M. NOTTINGHAM, Head, Biotechnology Division, The Meat Industry Research Institute of New Zealand (Inc.), P.O. Box 617, Hamilton, New Zealand

Dr J.C. OLSON, JR., Director, Division of Microbiology, Bureau of Foods, Food and Drug Administration, U.S. Department of Health, Education and Welfare, 200 C Street S.W., Washington, D.C. 20204, USA

Dr H. PIVNICK, Director, Bureau of Microbial Hazards, Food Directorate, Health Protection Branch, Health and Welfare Canada, Tunney's Pasture, Ottawa, Ontario, Canada, K1A 0L2

Dr FERNANDO QUEVEDO, Head, Food Microbiology and Hygiene Unit, Pan American Zoonoses Centre, Casilla 3092, 1000 Buenos Aires, Argentina

Dr J.H. SILLIKER, President, Silliker Laboratories, 1139 Dominguez St., Suite I, Carson, California 90746, USA

Mr BENT SIMONSEN, Chief Microbiologist, Danish Meat Products Laboratory, Howitzvej 13, Copenhagen F, Denmark

Professor H.J. SINELL, Director, Institute of Food Hygiene, Free University of Berlin, 1 Berlin 33, Bitterstrasse 8–12, Germany

Dr M. VAN SCHOTHORST, Head, Laboratory for Zoonoses and Food Microbiology, National Institute of Public Health, P.O. Box 1, Bilthoven, The Netherlands

MEMBERS (pre-1977, but active in preparation of this text)

Dr H.E. BAUMAN, Vice-President, Science and Technology, Research and Engineering, The Pillsbury Company, 608 Avenue S., Minneapolis, Minnesota 59402, USA

Professor C.E. DOLMAN, Professor Emeritus, Department of Microbiology, University of British Columbia, Vancouver, British Columbia, Canada V6T 1W5

Mr R. PAUL ELLIOTT, Consultant in Food Microbiology, 1095 Lariat Lane, Pebble Beach, California 93953, USA (Formerly Chief, Microbiology Staff, Meat and Poultry Inspection Program, Animal and Plant Health Inspection Service, U.S. Department of Agriculture, Washington, D.C. 20250, USA)

Dr H.E. GORESLINE, Consultant in Food Microbiology, Laakachtal 4, 2572 Kaumberg, Austria (Formerly with the Joint FAO/IAEA Division for Atomic Energy in Food and Agriculture, International Atomic Energy Agency, Kaerntnerring, Vienna 1, Austria)

Professor H. IIDA, Department of Microbiology, Hokkaido University, School of Medicine, North 15, West 7, Sapporo, Hokkaido, Japan

Professor M. INGRAM, Ex-Director, Meat Research Institute, Langford Nr. Bristol, BS18 7DY, England

Dr G.K. MORRIS, Chief, Epidemiologic Investigations, Laboratory Branch, Center for Disease Control, Department of Health, Education and Welfare, 1600 Clifton Rd., Bldg. 1, Rm. B 383, Atlanta, Georgia 30333, USA

Professor D.A.A. Mossel, Chairman, Food Bacteriology, Department of Food Science, School of Veterinary Medicine, Biltstraat 172, Utrecht, The Netherlands

Dr F.S. Thatcher (*Chairman 1962–73*), R.R.3, Merrickville, Ontario, Canada, K0G 1N0 (Formerly Chief, Division of Microbiology, Health Protection Branch, Health and Welfare Canada, Ottawa, Ontario, Canada)

LATIN-AMERICAN SUBCOMMISSION (current, 1977)

Principal members

Professor Josefina Gomez-Ruiz (*Chairwoman*), Central University of Venezuela, Apartado 50259, Caracas, Venezuela

Dra Silvia Mendoza G. (*Secretary*), Division of Biological Sciences, Department of Bioengineering, Simon Bolivar University, Apartado 5354, Caracas, Venezuela

Professor Nenufar Sosa de Caruso, Director, Institute of Food Industry, Veterinary Faculty, University of Montevideo, Casilla de Correo 753, Montevideo, Uruguay

Dr Fernando Quevedo, Head, Food Microbiology and Hygiene Unit, Pan American Zoonoses Centre, Casilla 3092, 1000 Buenos Aires, Argentina

Dr Sebastiao Timo Iaria, Faculty of Hygiene and Public Health, University of Sao Paulo, Sao Paulo S.P., Brazil

Observer members

Dra Ethel G.V. Amato de Lagarde, Instituto Nacional de Microbiologia 'Carlos G. Malbrán,' Avda Velez Sársfield 563, Buenos Aires, Argentina

Dra Elvira Regús de Valera, Calle 4, No. 15, Urbanizacion Miramar, Kilométro $8^1/_2$, Carretera Sanchez, Sto. Domingo, República Dominicana

Dr Mauro Faber de Freitas Leitao, Head, Department of Food Microbiology, Instituto de Tecnologia de Alimentos, Caixa Postal 139, 13.100 Campinas, Sao Paulo, Brazil

Dr Hernán Puerta Cardona, Chairman, Food Hygiene Section, Escuela Nacional de Salud Pública, Universidad de Antioquia, Apartado Aéreo 51922, Medellin, Colombia

Dra Alina Ratto, Universidad Nacional, Mayor de San Marcos, Apartado 5653, Lima, Peru

BALKAN AND DANUBIAN SUBCOMMISSION (current, 1977)

Professor Dr J. Takács (*Chairman*), Director, Institute of Food Hygiene, University of Veterinary Sciences, 1400 Budapest, P.O. Box 2, Hungary

Dr Milica Kalember-Radosavljević (*Secretary*), Food Bacteriologist, Military Academy, Institute of Hygiene, 2 Pasterova Avenue, Belgrade, Yugoslavia

Dr VLADIMIR BARTL, Head, Hygiene Laboratories, Hygiene Station for Middle Czech Region, Safarikova 14, 120 00 Praha 2, Czechoslovakia

Dr DEAC CORNEL, Institutul de Igiena, Str. Pasteur 6, Cluj, Rumania

Professor Dr O. PRANDL, Director, Institute of Meat Hygiene and Veterinary Food Technology, Vienna III/40, Linke Bahngasse 11, Austria

Dr S. TZANNETIS, Faculty of Medicine, Department of Microbiology, National University of Athens, Athens 609, P.O. Box 1540, Greece

Dr FAUD YANC, Sehir Hifzissihha, Müessesesi, Sarachanebasi, Istanbul, Turkey

Professor Dr Z. ZACHARIEV, Institute for Veterinary Medicine, Boulevar Lenina 55, Sofia, Bulgaria

MIDDLE EAST-NORTH AFRICAN SUBCOMMISSION (current, 1977)

Professor REFAT HABLAS (*Chairman*), Bacterial Typing Project, Al-Azhar Faculty of Medicine, Nasr City, Cairo, Egypt

Dr HASSAN SIDAHMED (*Secretary*), Head, Department of Bacteriology, P.O. Box 287, National Health Laboratory, Khartoum, Sudan

Professor A. ALAOUI, Director of Institute for National Hygiene, Professor of Microbiology, Rabat School of Medicine, Rabat, Morocco

Dr ABDUL-KAREEM NASIR AL-DULAIMI, Head, Department of Food and Water Hygiene, Central Public Health Laboratories, Andulis Square, Alwia, Baghdad, Iraq

Mr I. KASHOULIS, Analyst, Government Laboratory, Ministry of Health, Nicosia, Cyprus

Professor EL-SAYED EL-MOSSALAMI, Head, Meat Hygiene Department, Faculty of Veterinary Medicine, Cairo University, Giza, Egypt

Mr YACOUB KHALID MOTAWA, Head of Food Control Laboratory, Microbiological Section, Preventative Health, Public Health Laboratory, Ministry of Health, Kuwait

Dr NEJI OTHMAN, Chief, Laboratory of Food Microbiology, Ministry of Health, National Institute of Nutrition, Tunis, Tunisia

CONSULTANTS FOR THIS BOOK

Mr G.G. ANDERSON, Assistant Director, Inspection Branch, Fisheries and Marine Service, Department of the Environment and Marine, Ottawa, Ontario, Canada, K1A 0H3

Dr H. BEERENS, Director, Centre for Technical Study and Research for the Food Industry (CERTIA), 369 Rue Jules Guesde, Flers-Bourg, 59650 Villeneuve D'Ascq, Lille, France

Dr D.F. BRAY, Director of Planning and Evaluation, Health Protection Branch, Health and Welfare Canada, Tunney's Pasture, Ottawa, Ontario, Canada, K1A 0L2

Professor IRVING BURR, Professor of Statistics, Department of Statistics, Purdue University, Mathematical Sciences Building, Lafayette, Ind. 47907, USA

Dr E.F. DRION, Advisor on Experimental Design, Central Organization for Applied Scientific Research, TNO, P.O. Box 297, The Hague, The Netherlands

Mr I.E. ERDMAN, Chief, Evaluation Division, Bureau of Microbial Hazards, Health Protection Branch, Health and Welfare Canada, Tunney's Pasture, Ottawa, Ontario, Canada, K1A 0L2

Dr E.J. GANGAROSA, Deputy Chief, Bacterial Disease Branch, Center for Disease Control, U.S. Department of Health Education and Welfare, Atlanta, Georgia, USA

Dr A. HURST, Research Scientist, Microbiology Division, Health Protection Branch, Health and Welfare Canada, Tunney's Pasture, Ottawa, Ontario, Canada, K1A 0L2 (Past member of ICMSF)

Dr L. JOVANOVIĆ, Formerly Head, Bacteriology Department, Institute for Health Protection of Serbia, 15 Dr Jovana Subotica Street, Belgrade, Yugoslavia; now retired

Professor J. LISTON, Director, Institute for Food Science and Technology, College of Fisheries, University of Washington, Seattle, Wash. 98105, USA

Dr M.S. LOWENSTEIN, Chief, Enteric Diseases Section, Center for Disease Control, U.S. Department of Health Education and Welfare, Atlanta, Georgia 30333, USA

Mr D.A. LYON, Chief, Drug Statistics Services, Health Protection Branch, Health and Welfare Canada, Tower B, 355 River Road, Vanier, Ontario, Canada, K1A 1B8

Professor Z. MATYAS, Chief, Division of Communicable Diseases, World Health Organization, 1211 Geneva, Switzerland

Dr C.F. NIVEN, JR., Director of Research, Del Monte Research Center, 205 North Wiget Lane, Walnut Creek, California 94598, USA

Dr L. ORMAY, Chief, Department of Food Microbiology, Institute of Nutrition, Budapest IX, Gyali-ut 3/a, Hungary

Dr H. PIVNICK, Director, Bureau of Microbial Hazards, Food Directorate, Health and Welfare Canada, Tunney's Pasture, Ottawa, Ontario, Canada, K1A 0L2

Dr L. REINIUS, Food Hygienist, Veterinary Public Health, Division of Communicable Diseases, World Health Organization, 1211 Geneva 27, Switzerland

Dr J.M. SHEWAN, Chief, Microbiology Department, Torry Research Station, P.O. Box 31, 135 Abbey Road, Aberdeen AB9 8DG, Scotland

Dr J.H. SILLIKER, President, Silliker Laboratories, 1139 Dominguez St., Suite I, Carson, California, 90746, USA

Dr P. VASSILIADIS, Professor of Microbiology, Athens School of Hygiene, 196 Alexandras Avenue, Athens 602, Greece

Appendix 3

MEMBERS OF SUBCOMMITTEES ACTIVE IN THE PREPARATION OF THIS TEXT[1]

Disease Severity and Epidemiology
C.E. Dolman (*Chairman*), C. Cominazzini, O. Emberger, B.C. Hobbs, M.S. Lowenstein, H. Lundbeck, G. Mocquot, G.K. Morris, D.A.A. Mossel

Dried Foods
D.A.A. Mossel (*Chairman*), H.E. Bauman, D.S. Clark, J.H.B. Christian, O. Emberger, H.E. Goresline, J.H. Silliker

Editorial
M. Ingram (*Chairman*), D.F. Bray, D.S. Clark, R.P. Elliott, C.E. Dolman, F.S. Thatcher

Fish and Fishery Products
J.M. Shewan (*Chairman*), G.G. Anderson, J. Liston

Frozen Foods
K.H. Lewis (*Chairman*), H. Beerens, R.P. Elliott, I.E. Erdman, H. Lundbeck, S. Mendoza, N.P. Nefedjeva

Milk and Milk Products
G. Mocquot (*Chairman*), N.S. Caruso, C. Cominazzini, Z. Matyas, J.C. Olson, F. Quevedo, M. Šipka, F.S. Thatcher

Planning and Organization
F.S. Thatcher (*Chairman*), D.F. Bray, D.S. Clark

Raw and Processed Meats
B.C. Hobbs (*Chairman*: Raw Meats), B. Simonsen (*Chairman*: Processed Meats), J. Gomez-Ruiz, H. Iida, M. Ingram, C.E. Niven, J. Takács

Sampling Plans for Salmonellae
M. Ingram (*Chairman*), R.P. Elliott, G. Mocquot, C.F. Niven, J.C. Olson

Statistics
D.F. Bray (*Chairman*), I. Burr, E.F. Drion, D.A. Lyon

Shellfish
K.H. Lewis (*Chairman*), I.E. Erdmann, F.S. Thatcher

Vegetables
C.F. Niven (*Chairman*), Ph. V. Bartl, H.E. Bauman, J.H.B. Christian, Z. Matyas, G.K. Morris

1 Not all members were active throughout.

Appendix 4

CONTRIBUTORS TO THE SUSTAINING FUND OF ICMSF (updated to 1977)

American Can Co., America Lane, Greenwich, Conn. 06830, USA
Bacon and Meat Manufacturers Association, 1–2 Castle Lane, London, England, SW1E 6DU
Beatrice Foods Co., 120 South La Salle, Chicago, Ill. 60603, USA
Beecham Group Ltd., Beecham House, Great West Rd., Brentford, Middlesex, England
Brooke-Bond Oxo Ltd., Trojan Way, Purley Way, Croydon, England, CR9 9EH
Brown and Polson Ltd., Clay Gate House, Littleworth Rd., Esher, Surrey, England
Burns Foods Ltd., P.O. Box 1300, Calgary, Alta., Canada, T2P 2L4
Cadbury Schweppes Foods Ltd., Bournville, Birmingham, England
Cadbury Schweppes Powell Ltd., 1245 Sherbrooke St. W., Suite 1625, Montreal, Que., Canada, H3G 1G6
Campbell Soup Co. Ltd., 60 Birmingham St., Toronto, Ont., Canada, M8V 2B8
Canada Packers Ltd., 2211 St. Clair Ave. W., Toronto 9, Ont., Canada
Carlo Erba Institute for Therapeutic Research, Milan 20159, Italy
Central Alberta Dairy Pool, 5302 Gaetz Avenue, Red Deer, Alta., Canada
Centro Studi sull' Alimentazione, Gino Alfonso Sada, P. za Diaz 7-20123, Milan, Italy
Christie Brown and Co., Ltd., 2150 Lake Shore Blvd. W., Toronto 500, Ont., Canada, M8V 1A3
Coca-Cola Co., 310 North Ave., N.W., Atlanta, Ga., USA
CPC International Inc., International Plaza, Englewood Cliffs, N.J. 07632, USA
Del Monte International, 215 Fremont St., San Francisco, Cal. 94119, USA
Difco Laboratories, Detroit, Mich. 48232, USA
Distillers Co. Ltd., 21 St. James Square, London, S.W.1, England
Findus Ltd., Bjuv, Sweden
Frigoscandia Ltd., Fack S-215 01, Helsingborg, Sweden
General Foods Corporation, Technical Center, White Plains, N.Y. 10602, USA
General Foods Canada Ltd., Box 4019, Terminal A, Toronto, Ont., Canada
Gerber Products Co., 445 State St., Fremont, Mich. 49412, USA
H.J. Heinz Co. Ltd., Hayes Park, Hayes, Middlesex, England

Home Juice Co. Ltd., 175 Fenmar Dr., Weston, Ont., Canada
Horne & Pitfield Foods Ltd., 14550 112th Ave., P.O. Box 2266, Edmonton 15, Alta., Canada
Infant Formula Council, 64 Perimeter Center East, Atlanta, Ga. 30346, USA
International Union of Biological Sciences, 51 Bd. de Montmorency, 75016 Paris, France
ITT Continental Baking Co., P.O. Box 731, Rye, N.Y. 10580, USA
J. Sainsbury Ltd., Stamford House, Stamford Street, London, England, SE1 9LC
Jannock Corporation Ltd., P.O. Box 7, Montreal 101, Que., Canada
John Labatt Ltd., 451 Ridout St., London, Ont., Canada, N6A 2P6
Joseph Rank Ltd., Millcrat House, Eastcheap, London, E.C.3, England
Kellogg/Salada Canada Ltd., 6700 Finch Ave. W., Rexdale, Ont., Canada, M9W 5P2
Kraft Foods Ltd., Box 1673N, G.P.O., Melbourne, 3001, Australia
Langnese-Iglo Ltd., Hamburg, Germany
Maple Leaf Mills Ltd., P.O. Box 370, Station A, Toronto, Ont., Canada, M5W 1C7
Marks and Spencer Ltd., Michael House, Baker St., London, W.1, England
Mars Ltd., Dundee Rd., Trading Estate, Slough, Bucks., England
McCormick and Co. Inc., Baltimore, Md. 21202, USA
Milk Marketing Board, Thames Ditton, Surrey, England
Ministry of Health, Kuwait
RHM Research Ltd., Lincoln Road, High Wycombe, Bucks., H12 3QN, England
RJR Foods Inc., 4th and Main Sts., Winston Salem, N.C. 27102, USA
Reckitt and Colman Ltd., Carrow, Norwich, England, NR1 2DD
Ross Laboratories, 615 Cleveland Ave., Columbus, Ohio 43216, USA
Spillers Ltd., Old Charge House, Cannon St., London, E.C.4, England
Standard Brands Canada Ltd., 31 Airlie St., LaSalle, Que., Canada
Swift Canadian Co. Ltd., 2 Eva Rd., Etobicoke, Ont., Canada, M9C 4V5
Tate and Lyle Refineries Ltd., 21 Mincing Lane, London, England
Terme de Crodo, via Cristoforo Gluck 35, Milan 20125, Italy
Tesco Stores Ltd., Tesco House, Delamere Rd., Cheshunt, Waltham Cross, Herts., England
The Borden Co. Ltd., 1275 Lawrence Ave. E., Don Mills (Toronto), Ont., Canada
The J. Lyons Group of Companies, Cadby Hall, London, W14 09A, England
The Pillsbury Co., 311 Second St. S.E., Minneapolis, Minn. 55414, USA
The Quaker Oats Co., 617 West Main St., Barrington, Ill. 60010, USA
The Quaker Oats Co. Canada Ltd., Quaker Park, Peterborough, Ont., Canada
Unilever Ltd., Unilever House, Blackfriars, London, E.C.4, England
World Health Organization, Geneva, Switzerland

Appendix 5

A PROPOSED STATEMENT OF POLICY ON THE COMMERCIAL PROCESSING OF FOODS IN HERMETICALLY SEALED CONTAINERS

National Canners Association, 1971

1. Every canner (except meat and poultry canners) must register with the US FDA his name and place of business plus the location of each cannery he operates
2. All containers must be coded to show when and where the product was packed
3. A list of all low-acid products packed in each cannery must be filed, along with information on the processing equipment used
4. The processes to be used for each product must be filed, including container size, and times and temperatures
5. Suitable recording instruments must be employed on all retorts and cookers. The recorded data on each lot must be retained for at least five years
6. Period codes must be changed at least every four hours
7. In case of inadvertent underprocessing, the lot may be reprocessed, destroyed, or held pending evaluation of the safety aspects
8. The canner must report immediately to US FDA all instances of spoilage having potential health significance, if any of the product has entered distribution
9. Retort supervisors and operators must attend a certified school and complete a course on proper operation of the retort
10. Can seam inspectors must attend a certified school and receive a certificate upon completion of the course

Appendix 6

SANITATION OF SHELLFISH GROWING AREAS

The material in this Appendix appeared originally as Section C (Growing area survey and classification) and Appendix A (Bacteriological criteria for shucked oysters at the wholesale market level) of the manual 'National Shellfish Sanitation Program Manual of Operations: Part 1, Sanitation of Shellfish Growing Areas,' 1965 revision, edited by Leroy S. Houser, Sanitarian Director, and published by the US Department of Health Education and Welfare, Public Health Service.

GROWING AREA SURVEY AND CLASSIFICATION

1 *Sanitary surveys of growing areas*

A sanitary survey shall be made of each growing area prior to its approval by the State as a source of market shellfish or of shellfish to be used in a controlled purification or relaying operation. The sanitary quality of each area shall be reappraised at least biennially and, if necessary, a resurvey made. Ordinarily, resurveys will be much less comprehensive than the original survey since it will only be necessary to bring the original information up to date. Records of all original surveys and resurveys of growing areas shall be maintained by the State shellfish control agency, and shall be made available to Public Health Service review officers upon request.

Satisfactory compliance This item will be satisfied when

(a) A sanitary survey has been made of each growing area in the State prior to initial approval of interstate shipments of shellfish from that area. A comprehensive sanitary survey shall include an evaluation of all sources of actual or potential pollution on the estuary and its tributaries, and the distance of such sources from the growing areas; effectiveness and reliability of sewage treatment works; the presence of industrial wastes, pesticides, or radionuclides which would cause a public health hazard to the consumer of the shellfish; and the effect of wind, stream flow, and tidal currents in distributing polluting materials over the growing area.[1] The thoroughness with which each element

1 In making the sanitary survey, consideration should be given to the hydrographic and geographic characteristics of the estuary, the bacteriological quality of the growing area water and bottom sediments, and the presence and location of small sources of pollution, including boats, which might contribute fresh sewage to the area.

must be investigated varies greatly and will be determined by the specific conditions in each growing area.

(b) The factors influencing the sanitary quality of each approved shellfish growing area are reappraised at least biennially.[2] A complete resurvey should be made of each growing area in an approved category at least once every ten years; however, data from original surveys can be used when it is clear that such information is still valid.

(c) A file which contains all pertinent sanitary survey information, including the dates and results of preceding sanitary surveys is maintained by the State shellfish control agency for each classified shellfish area.

(d) The State agency having primary responsibility for this element of the national program develops a system for identification of growing area.

Public-health explanation The positive relationship between sewage pollution of shellfish growing areas and enteric disease has been demonstrated many times (5–10, 33–35). However, epidemiological investigations of shellfish-caused disease outbreaks have never established a direct numerical correlation between the bacteriological quality of water and the degree of hazard to health. Investigations made from 1914 to 1925 by the States and the Public Health Service – a period when disease outbreaks attributable to shellfish were more prevalent – indicated that typhoid fever or other enteric disease would not ordinarily be attributed to shellfish harvested from water in which not more than 50 per cent of the 1 cc portions of water examined were positive for coliforms,[3] provided the areas were not subject to direct contamination with small amounts of fresh sewage which would not ordinarily be revealed by the bacteriological examination.

Following the oyster-borne typhoid outbreak during the winter of 1924–5 in the United States (11), the national shellfish certification program was initiated by the States, the Public Health Service, and the shellfish industry. Water quality criteria were then stated as:

(a) The area is sufficiently removed from major sources of pollution so that the shellfish would not be subjected to faecal contamination in quantities which might be dangerous to the public health.

(b) The area is free from pollution by even small quantities of fresh sewage. The report emphasized that bacteriological examination does not, in itself, offer conclusive proof of the sanitary quality of an area.

(c) Bacteriological examination does not ordinarily show the presence of the coli-aerogenes group of bacteria in 1 cc dilutions of growing area water.

The reliability of this three-part standard for evaluating the safety of shellfish-producing areas is evidenced by the fact that no major outbreaks of typhoid fever or other enteric disease have been attributed to shellfish harvested from waters meeting the criteria since they were adopted in the United States in 1925. Similar water quality criteria have been in use in Canada with like results. The available epidemiological and laboratory evidence gives little idea as to the

2 The purpose of this reappraisal is to determine if there have been changes in stream flow, sewage treatment, populations, or other similar factors which might result in a change in the sanitary quality of the growing area. The amount of field work associated with such a reappraisal will depend upon the area under consideration and the magnitude of the changes which have taken place.

3 An MPN of approximately 70 per 100 ml

margin of safety, but it is probably considerable as indicated by the virtual absence of reported shellfish-caused enteric disease over a comparatively long period of time (10, 12, 13, 35, 37) from waters meeting this criteria.

The purpose of the sanitary survey is to identify and evaluate those factors influencing the sanitary quality of a growing area and which may include sources of pollution, potential or actual; the volume of dilution water; the effects of currents, winds, and tides in disseminating pollution over the growing areas; the bacterial quality of water and bottom sediments; die-out of polluting bacteria in the tributaries and the estuary; bottom configuration; and salinity and turbidity of the water. Sources of pollution include municipal sewage discharged into the estuary or inflowing rivers; sewage brought into the estuary by tides or currents; surface runoff from polluted areas; industrial wastes; and discharges from pleasure craft, fishing boats, naval vessels, and merchant shipping.

Bacteriological examination of the growing waters is an important component of the sanitary surveys. In many instances the bacteriological and related salinity data will also provide valuable information on the hydrographic characteristics of an area.[4,5]

Ideally, a large number of water samples for bacteriological examination should be collected at each station. However, in most instances this is not practical because of time and budget limitations, and accordingly only a limited number of samples can be collected. Therefore, sampling stations should be chosen which will provide a maximum of data, and which will be representative of the bacteriological quality of water in as wide an area as possible. Sample collection should be timed to represent the most unfavourable hydrographic and pollution conditions since shellfish respond rapidly to an increase in the number of bacteria or viruses in their environment (17–18, 38–40).

There is no specified minimum number of sampling stations, frequency of sampling, or total number of samples. Sampling results obtained over a period of several years can be used as a block of data, provided at least 15 samples have been collected from each of a representative number of stations along the line separating approved from restricted growing areas and there have been no adverse changes in hydrographic or sanitary conditions. Only occasional bacteriological samples are necessary from areas which are shown to be free from pollution.

4 Bacteria in an unfavourable environment die-out in such a way that following an initial lag period there is a large percentage decline during the first few days. Descriptions of studies on bacteria die-out have been published by Greenberg (14) and Pearson (15). Die-off has also been investigated by the Public Health Service Shellfish Sanitation Laboratory at Woods Hole, Mass., and Pensacola, Fla. Application of this principle may be helpful in predicting the quantity of pollution which will reach an area, and in establishing objective effluent quality criteria (16).

5 In connection with the evaluation of sampling results, it should be noted that the MPN determination is not a precise measure of the concentration of bacteria. Thus, in repeated sampling from waters having a uniform density of bacteria, varying MPN estimates will be obtained. The use of the tolerance factor 3.3 (applicable only to 5 tube decimal dilution MPNs) is one method of recognizing this variation. For example, in a body of water in which the median concentration of coliform bacteria is 70 per 100 ml, 95% of observed MPNs will be between 20 and 230 per 100 ml; i.e. $70/3.3 = 21$ and $70 \times 3.3 = 230$.

Experience with the shellfish certification program indicates a tendency to omit or deemphasize some components of the sanitary survey unless a central State file of all shellfish sanitary surveys, reappraisals, and resurveys is maintained. This is particularly true where responsibility for shellfish sanitation is divided between two or more State agencies. Maintenance of a central State file for all shellfish sanitary survey information will also simplify the endorsement appraisal of State programs by the Public Health Service and will help prevent loss of old data which may be useful in evaluating the sanitary quality of an area.

Periodic reappraisals of the sanitary quality of shellfish-producing areas are necessary to determine that environmental conditions are such that the original conclusions are still valid. A *resurvey* should be made within one year if the *reappraisal* shows a significant detrimental change.

2 *Classification of growing areas*

All actual and potential growing waters shall be classified as to their public health suitability for the harvesting of market shellfish. Classification criteria are described in sections C-3, C-4, C-5, C-6, and C-7 of this manual. Except in emergency, any upward revision of an area classification shall be preceded by a sanitary survey, resurvey, or reappraisal. A written analysis of the data justifying the reclassification shall be made a part of the area file.

Satisfactory compliance This item will be satisfied when

(a) All actual and potential growing waters in the State are correctly designated with one of the following classifications on the basis of sanitary survey information: *Approved*; *conditionally approved*; *restricted*; or *prohibited*.[6*]

(b) Area classifications are revised whenever warranted by survey data.

(c) Classifications are not revised upward without at least a file review, and there is a written record of such review in the area file maintained by the State shellfish control agency.

(d) All actual and potential growing areas which have not been subjected to sanitary surveys shall be automatically classified as *prohibited*.

Public-health explanation The probable presence or absence of pathogenic organisms in shellfish waters is of the greatest importance in deciding how shellfish obtained from an area may be used. All actual and potential growing waters should thus be classified according to the information developed in the sanitary survey. Classification should not be revised upward without careful consideration of available data. Areas should be reclassified whenever warranted by existing data. A written justification for the reclassification simplifies Public Health Service appraisal of State programs.

A hypothetical use of the four recognized area classifications is shown in Figure 7. This idealized situation depicts an estuary receiving sewage from two cities, A and B. City A has complete sewage treatment including chlorination of effluent. City B has no sewage treatment. The estuary has been divided into five areas, designated by roman numerals, on the basis of sanitary survey information:

6 Closures may also be based on the presence of Marine Toxins or other toxic materials.

* States may use other terminology in describing area classifications: provided, that the classification terms used are consistent with the intent and meaning of the words 'approved,' 'conditionally approved,' 'restricted,' or 'prohibited.'

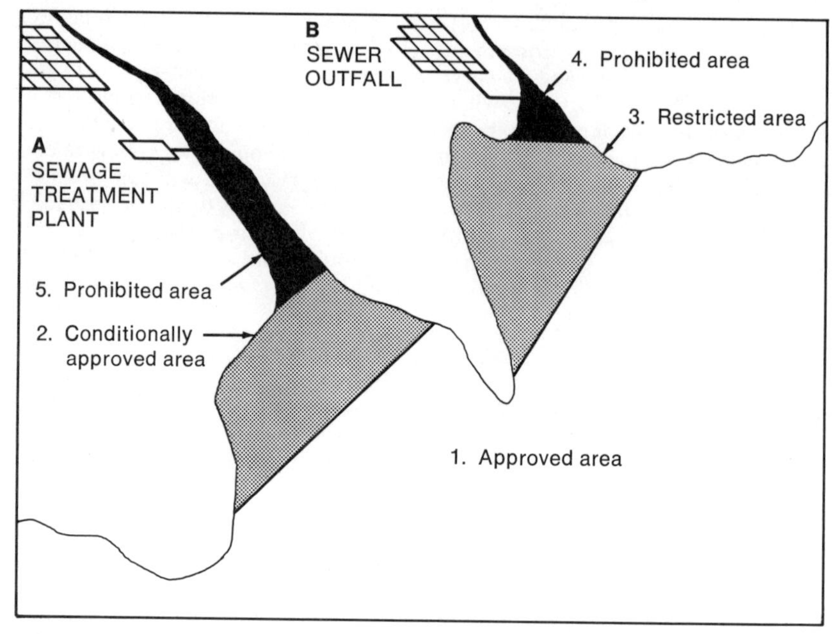

Figure 7

Approved

Area 1 The sanitary survey indicates that sewage from cities A and B (even with the A sewage plant not functioning) would not reach this area in such concentration as to constitute a public-health hazard. The median coliform MPN of the water is less than 70/100 ml. The sanitary quality of the area is independent of sewage treatment at city A.

Conditionally approved

Area 2 This area is of the same sanitary quality as area 1: however, the quality varies with the effectiveness of sewage treatment at city A. This area would probably be classified *prohibited* if city A had not provided sewage treatment.

Restricted

Area 3 Sewage from B reaches this area, and the median coliform MPN of water is between 70 and 700 per 100 ml. Shellfish may be used only under specified conditions.

Prohibited

Area 4 Direct harvesting from this area is prohibited because of raw sewage from B. The median coliform MPN of water may exceed 700/100 ml.

Area 5 Direct harvesting from this area is prohibited because of possible failure of the sewage treatment plant. Closure is based on need for a safety factor rather than coliform content of water or amount of dilution water.

3 *Approved areas*

Growing areas may be designated as *approved* when: (a) the sanitary survey indicates that pathogenic microorganisms, radionuclides, and/or harmful industrial wastes do not reach the area in dangerous concentration, and (b) this is verified by laboratory findings whenever the sanitary survey indicates the need. Shellfish may be taken from such areas for direct marketing.

Satisfactory compliance This item will be satisfied when the three following criteria are met:
(a) The area is not so contaminated with faecal material that consumption of the shellfish might be hazardous.
(b) The area is not so contaminated with radionuclides or industrial wastes that consumption of the shellfish might be hazardous.
(c) The coliform median MPN of the water does not exceed 70 per 100 ml, and not more than 10 per cent of the samples ordinarily exceed an MPN of 230 per 100 ml, for a 5-tube decimal dilution test (or 330 per 100 ml, where the 3-tube decimal dilution test is used) in those portions of the area most probably exposed to faecal contamination during the most unfavourable hydrographic and pollution conditions. (Note: This concentration might be exceeded if less than 8 million cubic feet of a coliform-free dilution water are available for each population equivalent (*coliform*) of sewage reaching the area). The foregoing limits need not be applied if it can be shown by detailed study that the coliforms are not of direct faecal origin and do not indicate a public health hazard (11, 12).[7]

Public-health explanation A review of epidemiological investigations of disease outbreaks attributable to the consumption of raw shellfish reveals that two general situations prevail insofar as pollution of growing or storage areas is concerned.
(1) Gross sewage contamination of a growing or wet storage area. (A report of a 1910 outbreak of typhoid fever involving 41 persons notes that raw sewage from a city with a population of 30,000 was discharged only a few hundred feet away from clam beds and floats (19, 20). In 1947 a case of typhoid fever was attributed to clams harvested 200 yards from the outlet of a municipal sewage treatment plant (21). In the latter case, the coliform MPN of the harbour water exceeded 12,000 per 100 ml, and the area had been posted as closed to shellfish harvesting.)
(2) Chance contamination of a growing or wet storage area by fresh faecal material which may not be diffused throughout the entire area (6, 8, 9, 11) and therefore not readily detectable by ordinary bacteriological procedures. The possibility of chance contamination was noted by Dr Gurion in his report on a 1902 typhoid outbreak. He is quoted in Public Health Bulletin No. 86, 'there is a zone of pollution established by the mere fact of the existence of a populated city

7 This MPN value is based on a typical ratio of coliforms to pathogens and would not be applicable to any situation in which an abnormally large number of pathogens might be present. Consideration must also be given to the possible presence of industrial or agricultural wastes in which there is an atypical coliform to pathogen ratio (22).

upon the banks of a stream or tidal estuary which makes the laying down of oysters and clams in these waters a pernicious custom if persisted in, because it renders those articles of food dangerous at times, and always suspicious.' The 1956 outbreak of infectious hepatitis in Sweden (691 cases) attributed to oysters which were contaminated in a wet storage area is an example of such contamination (8). Similarly in 1939, 87 cases of typhoid were attributed to faecal contamination of a storage area by a typhoid carrier (6).

It is well established that shellfish from water having a median coliform MPN not exceeding 70 per 100 ml,[8] and which is also protected against chance contamination with faecal material, will not be involved in the spread of disease which can be attributed to initial contamination of the shellfish. This is not surprising since a water MPN of 70/100 ml is equivalent to a dilution ratio of about 8 million cubic feet of coliform-free water per day for the faecal material from each person contributing sewage to the area. This tremendous volume of water is available in shellfish growing areas through tidal action which is constantly bringing unpolluted water into the area.

Areas which are approved for direct market harvesting of shellfish which will be eaten raw must necessarily meet one general test; i.e., sewage reaching the growing area must be so treated, diluted, or aged that it will be of negligible public-health significance. This implies an element of time and distance to permit the mixing of the sewage or faecal material with the very large volume of diluting water and for a major portion of the microorganisms to die out. Studies of the natural die-off of microorganisms in an unfavourable marine environment have been summarized by Greenberg (14).

The effectiveness of sewage treatment processes must be considered in evaluating the sanitary quality of a growing area since the bacterial and viral content of the effluent will be determined by the degree of treatment which is obtained (2, 41–43). The results of bacteriological sampling must also be correlated with sewage treatment plant operation, and evaluated in terms of the minimum treatment which can be expected with a realization of the possibility of malfunctioning, overloading, or poor operation.

The presence of radionuclides in growing area waters may also have public-health significance since shellfish, along with other marine organisms, have the ability to concentrate such materials (23–26). The degree to which radioisotopes will be concentrated depends upon the species of shellfish and the specific radioisotope. For example, it has been reported that the Eastern oyster has a concentration factor of 17,000 for Zn^{65} whereas the concentration factor in soft tissues for Sr^{80} is approximately unity (23, 25). The distribution of the radioisotope in the shellfish and the biological half-life are also variable. Sources of radioactive materials include fall-out, industrial wastes, and nuclear reactors. Limiting maximum permissible concentrations of radioactive materials expressed in terms of specific radioisotopes and unidentified mixtures in water and food have been established (27–28). The current standard should be consulted in evaluating the public-health significance of detected radioactivity in market shellfish.

The bacterial quality of active shellfish will ordinarily be directly propor-

8 There is a third general consideration in which shellfish may be contaminated through mishandling. This is not related to growing area sanitation.

tional to the bacterial quality of the water in which they grew; however, considerable variation in individual determinations may be expected. The coliform MPNs of the shellfish usually exceed those of the overlying water because shellfish filter large quantities of water to obtain food, thereby concentrating the suspended bacteria. This relationship will depend upon the shellfish species, water temperature, presence of certain chemicals, and varying capabilities of the individual animals.

4 *Conditionally approved areas*

The suitability of some areas for harvesting shellfish for direct marketing is dependent upon the attainment of an established performance standard by sewage treatment works discharging effluent, directly or indirectly, to the area. In other cases the sanitary quality of an area may be effected by seasonal population, or sporadic use of a dock or harbour facility. Such areas may be classified as *conditionally approved*.

State shellfish control agencies shall establish *conditionally approved* areas only when satisfied that (a) all necessary measures have been taken to ensure that performance standards will be met, and (b) precautions have been taken to assure that shellfish will not be marketed from the areas subsequent to any failure to meet the performance standards and before the shellfish can purify themselves of polluting microorganisms.

Satisfactory compliance This item will be satisfied when

(a) The water quality requirements for an *approved* area are met at all times while the area is approved as a source of shellfish for direct marketing.

(b) An operating procedure for each *conditionally approved* area is developed jointly by the State shellfish control agency, local agencies, including those responsible for operation of sewerage systems, and the local shellfish industry. The operating procedure should be based on an evaluation of each of the potential sources of pollution which may affect the area. The procedure should establish performance standards, specify necessary safety devices and measures, and define inspection and check procedures. (These procedures are described in more detail in the following public-health explanation.)

(c) A closed safety zone is established between the *conditionally approved* area and the source of pollution to give the State agency time to stop shellfish harvesting if performance standards are not met.

(d) Boundaries of *conditionally approved* areas are so marked as to be readily identified by harvesters.

(e) Critical sewerage system units are so designed, constructed, and maintained that the chances of failure to meet the established performance standards due to mechanical failure or overloading are minimized.

(f) There is a complete understanding of the purpose of the *conditionally approved* classification by all parties concerned, including the shellfish industry. Successful functioning of the concept is dependent upon the wholehearted co-operation of all interested parties. If such co-operation is not assured, the State should not approve the area for direct harvesting of market shellfish.

(g) Any failure to meet the performance standards is immediately reported to the State shellfish control agency by telephone or messenger. In some instances States may find it desirable to delegate the authority for closing a *conditionally approved* area to a representative of the agency located in the immediate area.

(h) The State immediately closes *conditionally approved* areas to shellfish harvesting following a report that the performance standards have not been met. The area shall remain closed until the performance standards can again be met plus a length of time sufficient for the shellfish to purify themselves so that they will not be a hazard to the public health.
(i) The State shellfish control agency makes at least two evaluations during the shellfish harvesting season of each *conditionally approved* area including inspection of each critical unit of the sewerage system to determine the general mechanical condition of the equipment, the accuracy of recording devices, and the accuracy of reporting by the operating agency.
(j) It is discovered that failure to meet performance standards have not been reported by the operating agency, or if the performance standards are not met, the area will immediately revert to a *restricted* or *prohibited* classification.
(k) All data relating to the operation of a *conditionally approved* area, including operation of sewerage systems, are maintained in a file by the State shellfish control agency.

Public-health explanation The *conditionally approved* classification is designed primarily to protect shellfish growing areas in which the water quality might undergo a significant adverse change within a short period of time.[9] The change might result from overloading or mechanical failure of a sewage treatment plant, or bypassing of sewage at a lift station.

Water quality in many growing areas in the more densely populated sections of the country is, to some degree, dependent upon the operation of sewage treatment plants. For example, the boundaries of an approved shellfish area might be determined during a period when a tributary sewage treatment plant is operating at a satisfactory level. If there is some interruption in treatment, it follows that there will be some degradation in water quality in the growing area which may justify a relocation of the boundaries. The degree of relocation would depend upon such items as the distance between the pollution source and the growing area, hydrography, the amount of dilution water, and the amount of pollution.

The concept is also applicable to other situations in which there may be a rapid or seasonal change in water quality. Examples of such situations include:
(a) A growing area adjacent to a resort community. During the summer months the community might have a large population and this could have an adverse effect on water quality. However, during the winter when there are few people in the community, the water quality might improve sufficiently to allow approval of the area. In some States this is known as a seasonal closure.
(b) A protected harbour in a sparsely settled area might provide anchorage for a fishing fleet several months a year. When the fishing fleet is in, the harbour water would be of poor sanitary quality; however, during the remainder of the year the quality of the harbour water might be satisfactory. The area would be approved for shellfish harvesting only when the fishing fleet is not using the harbour.
(c) The water quality in an area fluctuates with the discharge of a major river. During periods of high runoff the area is polluted because of decreased flow time

9 A natural disaster may also cause many sewage treatment plants to be out of service for an extended period of time. The *conditionally approved* area concept is not ordinarily concerned with such emergency situations.

in the river. However, during periods of low runoff the area might be of satisfactory quality and thus be approved for shellfish harvesting.

The establishment of *conditionally approved* areas might be considered whenever the potential for sewage contamination is such that the limiting water quality criteria for an *approved* area might be exceeded in less than one week due to a failure of sewage treatment, or other situations as described above.

The first step in determining whether an area should be placed in the *conditionally approved* classification is the evaluation of the potential sources of pollution in terms of their effect on water quality in the area. Potential sources of pollution include the following:

1 Sewage treatment plants: (a) bypassing of all or part of sewage because of mechanical or power failure, hydraulic overloading, or treatment overloading; (b) reduced degree of treatment caused by operational difficulties or inadequate plant.

2 Sewage lift stations: (a) bypassing during periods of maximum flow due to inadequate capacity; (b) bypassing because of mechanical or power failure.

3 Interceptor sewers or underwater outfalls: (a) exfiltration due to faulty construction; (b) leakage due to damage.

4 Other sources of pollution: (a) sewage from merchant or naval vessels; (b) sewage from recreation use of area.

The second step in establishment of a *conditionally approved* area is the evaluation of each source of pollution in terms of the water quality standards to be maintained, and the formulation of performance standards for each installation having a significant effect on the sanitary quality of the area. Examples of performance standards might include:

1 Bacteriological quality of effiuent from sewage treatment plants. This might be stated in terms of chlorine residual if the bacteriological quality of the effluent can be positively related to chlorine residual. The following is an example of a performance standard (29): 'The median coliform MPN, in any one month, shall not exceed 500 per 100 ml, based on not less than 16 composite samples per month, and not more than 10 per cent of the samples shall have an MPN in excess of 10,000 per 100 ml. Determinations of the chlorine residual of the effluent should be made hourly and recorded in the permanent plant records.'

2 Total quantity of sewage which can be discharged from any given unit, or from a combination of units, without causing the basic water quality standards to be exceeded.

3 Amount of shipping in the area and the amount of sewage that can be expected.

Design criteria which may be useful in formulating an opinion on the quantity of sewage which can be discharged into an area without exceeding the desired water quality standards include: population equivalent (*coliform*) of sewage; predicted survival of coliforms in sea water, effectiveness of chlorination, and the total quantity of clean dilution water in an area. Results of many studies on the survival of bacteria in sea water have been summarized in *An Investigation of the Efficacy of Submarine Outfall Disposal of Sewage and Sludge*; Publication no. 14, California State Water Pollution Control Board, 1956.

The mechanical equipment at critical sewage treatment or pumping units should be such that interruptions will be minimized. Wherever possible, operations should be automatically recorded on charts. Examples of the requirements which might be imposed, depending upon the importance of the unit in terms of water quality, include:

1 Ample capacity for storm flows. (Storm water should ordinarily be excluded from the sanitary system.)

2 Standby equipment to ensure that treatment or pumping will not be interrupted because of damage to a single unit or to power failure.

3 Instrumentation of pumps and equipment to allow the regulatory agency to determine that performance standards have been met. Examples include: (a) recording scales to indicate rate of chlorine use. Chlorine flow can be integrated with hydraulic flow to establish a ratio; (b) liquid level recording gages in overflow channels of sewage treatment plants and wet wells of lift stations to indicate when overflow takes place. Charts should be dated and initialed by the operator. Gages should be calibrated so that discharge can be estimated; (c) automatic devices to warn of failure or malfunctioning at self-operated pumping stations or treatment plants.

4 The effect of storm sewage can be calculated by multiplying the total estimated flow by the observed coliform content. The result can be expressed in terms of population equivalents (*coliform*).

Design and operation of equipment should be such that closure provisions should not have to be invoked more than once per year under ordinary circumstances.

A closed safety area should be interposed between the *conditionally approved* area and the source of pollution. The size of such area should be based on the total time it would take for the operating agency to detect a failure, notify the State shellfish control agency, and for the latter agency to stop shellfish harvesting. It is recommended that the area be of such size that the flow time through the safety area be at least twice that required for the notification process to become effective. Due consideration should be given to the possibility that closure actions might be necessary on holidays or at night.

The type of marking which will be required for *conditionally approved* areas will vary from State to State depending upon the legal requirements for closing an area.

The length of time a *conditionally approved* area should be closed following a temporary closure will depend upon several factors, including the species of shellfish, water temperature, purification rates, presence of silt or other chemicals that might interfere with the physiological activity of the shellfish, and the degree of pollution of the area.

5 *Restricted areas*

An area may be classified as *restricted* when a sanitary survey indicates a limited degree of pollution which would make it unsafe to harvest the shellfish for direct marketing. Alternatively the States may classify such areas as prohibited. Shellfish from such areas may be marketed after purifying or relaying.

Satisfactory compliance This item will be satisfied when the following water quality criteria are met in areas designated by States as *restricted*.[10, 11]

10 It is not mandatory that States use this classification. Areas not meeting the *approved* classification may be closed to all harvesting for direct marketing.

11 Routine sanitary surveys and reappraisals of *restricted* areas shall be made on the same frequency as for *approved* areas.

(a) The area is so contaminated with faecal materials that direct consumption of the shellfish might be hazardous, and/or
(b) The area is not so contaminated with radionuclides or industrial wastes that consumption of the shellfish might be hazardous, and/or
(c) The coliform median MPN of the water does not exceed 700 per 100 ml and not more than 10 per cent of the samples exceed an MPN of 2300 per 100 ml in those portions of the areas most probably exposed to faecal contamination during the most unfavourable hydrographic and pollution conditions. (Note: this concentration might be exceeded if less than 800,000 cubic feet of a coliform-free dilution water are available for each population equivalent (*coliform*) of sewage reaching the area.)
(d) Shellfish from *restricted* areas are not marketed without controlled purification or relaying.

Public-health explanation In many instances it is difficult to draw a clear line of demarcation between polluted and non-polluted areas. In such instances the State may, at its option, classify areas of intermediate sanitary quality as *restricted* and authorize the use of the shellfish for relaying, or controlled purification.

6 *Prohibited areas*

An area shall be classified *prohibited* if the sanitary survey indicates that dangerous numbers of pathogenic microorganisms might reach an area. The taking of shellfish from such areas for direct marketing shall be prohibited. Relaying or other salvage operations shall be carefully supervised to ensure against polluted shellfish entering trade channels. Actual and potential growing areas which have not been subjected to sanitary surveys shall be automatically classified as *prohibited*.

Satisfactory compliance This item will be satisfied when:
(a) An area is classified as *prohibited* if a sanitary survey indicates either of the following degrees of pollution:

1 The area is so contaminated with radionuclides or industrial wastes that consumption of the shellfish might be hazardous and/or
2 The median coliform MPN of the water exceeds 700 per 100 ml or more than 10 per cent of the samples have a coliform MPN in excess of 2300 per 100 ml. (Note: this concentration might be reached if less than 800,000 cubic feet of a coliform-free dilution water are available for each population equivalent (*coliform*) of sewage reaching the area.)

(b) No market shellfish are taken from *prohibited* areas except by special permit.
(c) Coastal areas in which sanitary surveys have not been made shall be automatically classified as *prohibited*.

Public-health explanation Prevention of the interstate transport of shellfish containing sufficient numbers of pathogenic microorganisms to cause disease is a primary objective of the National Program. Therefore, areas containing dangerous concentrations of microorganisms of faecal origin, or areas which may be slightly contaminated with fresh faecal discharges, should not be approved as a source of shellfish for direct marketing.

7 *Closure of areas due to shellfish toxins*

The State shellfish control agency shall regularly collect and assay representative samples of shellfish from growing areas where shellfish toxins are likely to occur. If the paralytic shellfish poison content reaches 80 micrograms per 100 grams of the edible portions of raw shellfish meat, the area shall be closed to the taking of the species of shellfish in which the poison has been found.[12]

The quarantine shall remain in effect until such time as the State shellfish control agency is convinced the poison content of the shellfish involved is below the quarantine level.[13]

Satisfactory compliance This item will be satisfied when

(a) The State shellfish control agency collects and assays representative samples of shellfish for the presence of toxins from each suspected growing area during the harvesting season.

(b) A quarantine is imposed against the taking of shellfish when the concentration of paralytic shellfish poison equals or exceeds 80 micrograms per 100 grams of the edible portion of raw shellfish.

Public-health explanation In some areas paralytic poison is collected temporarily by bivalve shellfish from free-swimming, one-celled marine plants on which these shellfish feed. The plants flourish seasonally when water conditions are favourable.

Cases of paralytic poisoning, including several fatalities, resulting from poisonous shellfish have been reported from both the Atlantic and Pacific coasts. The minimum quantity of poison which will cause intoxication in a susceptible person is not known. Epidemiological investigations of paralytic shellfish poisoning in Canada have indicated that 200 to 600 micrograms of poison will produce symptoms in susceptible persons and a death has been attributed to the ingestion of a probable 480 micrograms of poison. Investigations indicate that lesser amounts of the poison have no deleterious effects on humans. Growing areas should be closed at a lower toxicity level to provide an adequate margin of safety since in many instances toxicity levels will change rapidly (30–31). It has also been shown that the heat treatment afforded in ordinary canning processes reduces the poison content of raw shellfish considerably.

A review of literature and research dealing with the source of the poison, the occurrence and distribution of poisonous shellfish, physiology and toxicology, characteristics of the poison, and prevention and control of poisoning has been prepared (32).

In Gulf coast areas, toxicity in shellfish has been associated (4, 44) with Red Tide outbreaks caused by mass bloomings of the toxic dinoflagellate, *Gymnodinium breve*. Toxic symptoms in mice suggest a type of *ciguatera* fish poisoning rather than symptoms of paralytic shellfish poisoning.

12 This value is based on the results of epidemiological investigations of outbreaks of paralytic shellfish poison in Canada in 1954 and 1957 (30, 31).

13 The provisions of this item apply only to shellfish which will be marketed as a fresh or frozen product as properly controlled heat processing will reduce the poison content of the shellfish.

BACTERIOLOGICAL CRITERIA FOR SHUCKED OYSTERS AT THE WHOLESALE MARKET LEVEL

The development of satisfactory bacteriological criteria for interstate shipments of oysters as received at the wholesale market level has been under consideration since 1950. At that time the Canadian Department of National Health and Welfare pointed out that most of the US shucked Eastern oysters sold in Canada had high coliform MPNs, high standard plate counts, or both (2). The Canadian experience with market standards for oysters was discussed at the 1956 National Shellfish Sanitation Workshop (2) and the Workshop adopted on an interim basis the following bacteriological standard for shucked Eastern oysters at the wholesale market level:

Class 1, Acceptable

Shucked oysters with a Most Probable Number (MPN) of coliform bacteria of not more than 16,000 per 100 ml, and/or a Standard Plate Count of not more than 50,000 per ml.

Class 2, Acceptable on Condition

Shucked oysters with a coliform MPN greater than 16,000 per 100 ml, but less than 160,000 per ml, and/or a Standard Plate Count greater than 50,000 per ml, but less than 1 million per ml. (The oysters will be accepted on the condition that the shellfish sanitation authority in the originating State make immediate investigation of the producer's plant and operations and submit a report of such investigations to the control agency in the market area. On the basis of this report the control agency in the market will reject or permit further shipments from the producer in question.)

Class 3, Rejectable

Shucked oysters with a coliform MPN of 160,000 or more per 100 ml, and/or a Standard Plate Count of 1 million or more per ml.

In establishing the above interim standards, the 1956 Workshop recognized the limitations of the coliform group as an index of quality in that it failed to reveal whether the shellfish had been harvested from polluted areas or had been exposed to contamination during handling and processing subsequent to removal from the water. A recommendation was made that investigations be conducted to evaluate the significance of other bacterial indices. The faecal coliform group was suggested as a possible substitute for the coliform indices.

In partial fulfilment of this suggestion, a report on an interstate co-operative study to evaluate bacteriological criteria for market oysters was presented at the 1958 Shellfish Sanitation Workshop (3). A feature of this report was the development and evaluation of a method for the estimation of faecal coliform organisms following a procedure originally developed by Hajna and Perry (45). Gross increases in coliform organisms were observed during normal acceptable commercial practices. The magnitude of changes in coliform organisms was of the same order as those observed in plate counts. The results clearly demonstrated the inadequacy of the coliform group as an indicator of the sanitary quality of shellfish. It was further concluded that the plate count was of equal significance in revealing chance contamination or violations of acceptable storage time and temperature. On the other hand, the results of the examinations for faecal coliform organisms revealed a much higher degree of stability as the shellfish proceeded through commercial channels and thus suggested the greater suitability of this parameter as an index of sanitary quality at the wholesale market level. After due consideration of the report, the 1958 Workshop changed the interim bacteriological standard for fresh and frozen shucked oysters at the wholesale market level to the following:

Satisfactory[14]

E. coli density of not more than 78 MPN per 100 ml of samples as indicated by production of gas in E.C. liquid broth media nor more than 100,000 total bacteria per ml on agar at 35°C will be acceptable without question. An *E. coli* content of 79 to 230 MPN per 100 ml of sample or a total bacteria count of 100,000 to 500,000 per ml will be acceptable in occasional samples. If these concentrations are found in two successive samples from the same packer or repacker, the State regulatory authority at the source will be requested to supply information to the receiving State concerning the status of operation of this packer or repacker.

Unsatisfactory[14]

E. coli content of more than 230 MPN per 100 ml of sample or a total bacteria count of more than 500,000 per ml will constitute an unsatisfactory sample and may be subject to rejection by the State shellfish regulatory authority. Future shipments to receiving markets by the shipper concerned will depend upon satisfactory operational reports by the shellfish regulatory authorities at the point of origin.

In adopting the above standards, the 1958 Workshop recommended that the co-operative studies conducted by city and State laboratories and the Public Health Service be continued.

The 1961 Workshop reviewed still more data collected by the collaborating agencies during the 1958–61 period (36) and after considerable deliberation agreed to continued use of the interim bacteriological standards arrived at by the 1958 Workshop.

The 1964 Workshop considered all bacteriological data available up to that time (17–19 Nov.), including data relative to *Crassostrea gigas*, and adopted the following standards on a permanent basis, versus the previous interim basis, as being applicable to all species of fresh and frozen oysters at the wholesale market level, provided they can be identified as having been produced under the general sanitary controls of the National Shellfish Sanitation Program.[15]

14 *E. coli*, defined as coliforms which will produce gas from E.C. medium within 48 hours at 44.5°C in a water bath, will be referred to as faecal coliforms.

15 The standards are not considered meaningful in the absence of such information.

Satisfactory

Faecal coliform density[16] of not more than 230 MPN per 100 grams and 35°C plate count[17] of not more than 500,000 per gram will be acceptable without question.

Conditional

Faecal coliform density of more than 230 MPN per 100 grams and/or 35°C plate count of more than 500,000 per gram will constitute a conditional sample and may be subject to rejection by the State shellfish regulatory authority. If these concentrations are found in two successive samples from the same shipper, the State regulatory authority at the source will be requested to supply information to the receiving State concerning the status of operation of this shipper. Future shipments to receiving markets by the shipper concerned will depend upon satisfactory operational reports by the shellfish regulatory authorities at the point of origin.

In establishing the above bacteriological standards the 1964 Workshop took cognizance of the fact that no known health hazard was involved in consuming oysters meeting the standard; that oysters produced in the Gulf Coast States with warmer growing waters, could meet the standard if harvested, processed, and distributed according to the National Shellfish Sanitation Program requirements, and that the oysters harvested were from 'approved' growing areas complying with the standards for growing areas established in part I of the PHS Publication no. 33.

REFERENCES

1 Jensen, E.T. 1955. The 1954 National Conference on Shellfish Sanitation. Public Health Repts., vol. 70, no. 9
2 Proceedings 1956 Shellfish Sanitation Workshop, mimeographed, Public Health Service, 1956
3 Proceedings 1958 Shellfish Sanitation Workshop, lithographed, Public Health Service, 1958
4 McFarren, E.F. Mimeograph 1–14–63. Available from P.H.S. Shellfish Sanitation Branch
5 Fisher, L.M., Report of the Committee of the Public Health Engineering Section of the American Public Health Association, Am. J. Public Health. *27*, 180, Suppl. March 1937
6 Old, H.N. and Gill, S.L. 1940. A typhoid fever epidemic caused by carrier bootlegging oysters. Am. J. Public Health. *30*, 633
7 Hart, J.C. 1945. Typhoid fever from clams. Connecticut Health Bull.

16 Faecal coliform organisms are those which, on transfer to E.C. medium from gas positive presumptive broth tubes show production of gas after incubation in a water bath at 44.5°C ± 0.2°C for 24 hours. Where air incubation is at 45.5°C ± 0.2°C comparative tests must be made to determine comparable time of incubation.

17 Plate count is the number of bacteria determined by the 'Standard Plate Count: procedure for shellfish described in the APHA Recommended Procedures for the Bacteriological Examination of Sea Water and Shellfish.'

8 Roos, Bertil. 1956. Hepatitis epidemic conveyed by oysters. Svenska Läkurtidningen. *53*, 989 (translation available from the Public Health Service)
9 Lindberg-Bromnn, Ann Muri. 1956. Clinical observations in the so-called oyster hepatitis. Svenska Läkurtidningen. *53*, 1003 (translation available from the Public Health Service)
10 Meyers, K.F. 1953. Medical progress – food poisoning. New Eng. J. Med. *249*, 765, 804, and 843
11 Lumsden, L.L., Hasseltine, H.E., Leak, J.P., and Veldee, M.V. 1925. A typhoid fever epidemic caused by oyster-borne infection. Public Health Reports, suppl. no. 50
12 A report on the public health aspects of clamming in Raritan Bay, Public Health Service, reissued June 1954
13 Dack, G.M. 1964. Food poisoning. 3rd ed. Chicago: University of Chicago Press
14 Greenberg, Arnold E. 1956. Survival of enteric organisms in sea water. Public Health Repts. *71*, no. 1
15 An investigation of the efficacy of submarine outfall disposal of sewage and sludge, publication no. 14, California State Water Pollution Control Board, 1956
16 Harris, Eugene K. 1958. On the probability of survival of bacteria in sea water, Biometrics
17 Wood, P.C. 1957. Factors affecting the pollution and self-purification of molluscan shellfish, Extrait du Journal du Conseil International Pour l'Exploration de la Mer, XXII, no. 2
18 Arcisz, William and Kelly, C.B. 1955. Self-purification of the soft clam, *Mya arenaria*. Public Health Repts. *70*, 605
19 Investigation of pollution of tidal waters of Maryland and Virginia. Public Health Bull. no. 74, 1916
20 Investigation of the pollution of certain tidal waters of New Jersey, New York and Delaware. Public Health Bull. no. 86, 1917
21 Mood, Eric W. 1948. First typhoid case in seven years. Monthly Rept. of the New Haven, Conn., Department of Health
22 Bidwell, Milton H. and Kelley, C.B. 1950. Ducks and shellfish sanitation, Am. J. Public Health, *40*, 8
23 Effects of atomic radiation on oceanography and fisheries. Publ. no. 551, Nat. Acad. Sci., Nat. Res. Council, 1957
24 Gong, J.K. *et al*. 1957. Uptake of fission products and neutron-induced radionuclides by the clam. Proc. Soc. Exp. Biol. Med. *95*, 451
25 Studies of the fate of certain radionuclides in estaurine and other aquatic environments. Public Health Service Publ. no. 999-R-3
26 Welsa, H.V. and Shipman, W.H. 1957. Biological concentration by killer clams of cobalt-60 from radioactive fallout. Science, *125*
27 Title 10, Part 20, Code of Federal regulations
28 Maximum permissible body burdens and maximum permissible concentrations of radionuclides in air and in water for occupational exposure. Nat. Bureau of Standards Handbook 69, 1959
29 Water quality survey of Hampton Roads shellfish areas. 1950. Virginia State Department of Health and US Public Health Service
30 Tennant, A.D., Neubert, J., and Corbeil, H.E. 1955. An outbreak of paralytic shellfish poisoning. Can. Med. Assoc. J. *72*, 436
31 Proceedings 1957 Conference on Paralytic Shellfish Poison, mimeographed, Public Health Service, 1958
32 McFarren, E.F. *et. al*. 1960. Public health significance of paralytic shellfish poison – advances in food research, *10*
33 Ringe, Mila E., Clem, David J., Linkner, Robert E., and Sherman, Leslie K. A case study on the transmission of infectious hepatitis by raw clams, published by US Department of Health, Education and Welfare, Public Health Service

34 Mason, James O. and McLean, W.R. 1962. Infectious hepatitis traced to the consumption of raw oysters. Am. J. Hygiene. *75*, no. 1

35 Communicable disease center hepatitis surveillance, Rept. no. 18, 1964, and Rept. no. 19, 1964. US Department of Health, Education and Welfare, Public Health Service

36 Proceedings 1961 Shellfish Sanitation Workshop, lithographed, Public Health Service, 1962

37 Communicable disease center hepatitis surveillance, Rept. no. 5, 1961, and Rept. no. 6, 1961. US Department of Health, Education and Welfare, Public Health Service

38 Metcalf, T.G. and Stiles, W.C. 1965. The accumulation of the enteric viruses by the oysters, *Crassostrea virginica*. J. Infect. Dis. *115*, 68

39 Hedstrom, C.E. and Lycke, E. An experimental study on oysters, Am. J. Hygiene. *79*, 143

40 Crovari, Piero. 1958. Some observations on the depuration of mussels infected with poliomyelitis virus, *Iqiene Moderna, 51*, 22. Translation available from PHS Shellfish Sanitation Branch

41 Kabler, Paul. 1959. Removal of pathogenic microorganisms by sewage treatment processes. Sewage and Industrial Wastes, *31*, 1373

42 Kelly, Salley and Sanderson, W.W. 1959. The effect of sewage treatment on viruses. Sewage and Industrial Wastes, *31*, 683

43 Clarke, Norman A. and Kabler, Paul W. 1964. Human enteric viruses in sewage. Health Lab. Sci. *1*, 44

44 Eldred, B., Steidinger, K., and Williams, J. 1964. Preliminary studies of the relation of *gymnodinum breve* counts to shellfish toxicity. A collection of data in reference to red tide outbreaks during 1963. Reproduced by the Marine Laboratory of the Florida Board of Conservation, St Petersburg, Florida

45 Hajna, A.A. and Perry, C.A. 1954. Comparative study of presumptive and confirmatory media for bacteria of the coliform group and for fecal streptococci. Am. J. Public Health, *33*, 550

GLOSSARY AND REFERENCES

Glossary

Acceptance sampling
The application of a predetermined sampling plan to a lot of a product to decide whether or not the lot meets defined criteria of acceptance.

AQL
Acceptance quality level. The percentage of units within a lot that must be free of a particular defect for the lot to be acceptable.

Aliquot
The portion of food, derived by weighing and dilution, that is inoculated into a container of bacteriological medium in accordance with a specified method. For example, from a 1 pound sample unit, a 25 g portion may be blended with 225 ml of diluent to give a 1:10 dilution. A 1 ml portion of this suspension would be an aliquot of 0.1 g.

Attributes plan
A sampling plan in which each selected sample unit is classified according to the quality characteristics of the product and in which there are only two or three grades of quality, e.g.: acceptable, defective; absent, present; acceptable, marginally acceptable, defective; low count, medium count, high count.

a_w
Water activity – a term expressing a general measure of the availability of water to microorganisms, defined as p/p_0 where p and p_0 are, respectively, the vapour pressures of the microbial substrate (food, solution, or microbiological medium) and of pure water. $p/p_0 \times 100$ is also the relative humidity of the atmosphere in equilibrium with the substrate. Thus, pure water has an a_w of 1.00, and is in equilibrium with an atmosphere of 100% relative humidity. The a_w of a solution (or food or bacterial medium) is invariably less than 1.00. To grow, microorganisms have specific cardinal requirements for a_w; hence, growth and/or spoilage in a food will be conditioned by the a_w of the food.

c
The maximum allowable number of defective sample units. When more than this number are found, the lot is rejected.

Case
A set of circumstances related to the nature and treatment of a food, categorized into 15 such sets (cases 1–15), which influence the anticipated hazard from the presence of specified bacterial species or groups within a food (see Table 5, page 29).

Coliforms
A multi-generic group of lactose-positive bacteria of the family Enterobacteriaceae, often further subdivided by definition into the three groups, coliforms, faecal coliforms, and

Escherichia coli, these being widely accepted as having progressively greater likelihood of being derived from a faecal source. (See Thatcher and Clark, 1968, for specific definitions and determinative methods.)

Consignment
A quantity, large or small, of food destined, in commerce, to a particular recipient, usually consisting of multiple containers of food from one or more lots.

Consumer risk
The probability that a lot will be accepted on the basis of the observed sample values when in reality the lot as a whole is substandard relative to the stated criteria of acceptability.

DMC
Direct Microscopic Count. A method for estimating the number of bacteria per gram of food by direct microscopic examination of a stained slide prepared from an aqueous suspension of the food.

Frame
That portion of the consignment from which the sample units are drawn. Ideally it should be the whole lot, in practice it is the accessible portion of the lot.

GCP
Good Commercial Practice. A general term defining conditions of GMP combined with acceptable conditions of distribution and storage in international commerce.

GMP
Good Manufacturing Practice. Those procedures in a food processing plant which consistently yield products of acceptable microbial quality, suitably monitored by laboratory tests. GMP is usually described in a code defining processes, equipment, plant layout, sanitation, hygiene, and laboratory tests; but as used here, does not apply to any particular code now in use.

Hazard
The likelihood based on experience that a food containing particular organisms will (i) cause disease when consumed; (ii) become organoleptically unacceptable (spoiled) in a period shorter than the normal shelf-life for such a food.

Indicator
Historically, an organism itself non-pathogenic, but often associated with pathogens, used to portray a risk of the presence of pathogens for which feasible methods of detection were not generally available. This usage is currently expanded to denote groups or species of organisms whose presence in a food reveal exposure to conditions that might introduce hazardous organisms and/or allow their growth. Specific indicators are now used to reveal excessively contaminated raw materials, unsanitary manufacturing practice, contamination from faecal, naso-pharyngeal, or suppurative sources, unsuitable time-temperature conditions of storage, failure of a process. The indicator groups and their methods of enumeration are described in Thatcher and Clark (1968).

Lot
A lot, in the commercial sense, is a quantity of food supposedly produced under identical conditions, all packages of which would normally bear a lot number that identifies the production during a particular time interval, and usually from a particular 'line,' retort or other critical processing unit (see page 9). Statistically, a lot is considered as a collection of sample units of a product from which a sample is to be drawn to determine acceptability of the lot.

m
A microbiological criterion which, in a 2-class plan separates good quality from defective quality; or, in a 3-class plan separates good quality from marginally acceptable quality. In general *m* represents an acceptable level and values above it are marginally acceptable or unacceptable.

M
A microbiological criterion which, in a 3-class plan, separates marginally acceptable quality from defective quality. Values at or above *M* are unacceptable.

Microbiological criterion
A microbiological value (e.g., number per g) established by use of defined procedures and applied in acceptance sampling of food (see *m* and *M*)

Microbiological limit
A microbiological criterion recommended by an authoritative body for adoption in specific regions but *not* incorporated into law.

Microbiological purchasing specification
The microbiological criterion or criteria conditional to acceptance of a specific food or food ingredient by a food manufacturer or other private or public purchasing agency.

Microbiological standard
A microbiological criterion in a law or regulation controlling foods produced, processed, or stored in, or imported into the area of jurisdiction of a regulatory agency.

Moderate hazard
The hazard associated with consumption of a food containing a pathogen or toxin which to people in normal health would usually cause a short-lived disease not critically severe in its manifestations and normally without untoward sequellae; for example, salmonellosis, staphylococcal food-poisoning.

MPN
The estimated number (per ml or per 100 ml) of the test organism present in a sample unit, based upon its presence or absence in replicate aliquots prepared by decimal dilution (see also Thatcher and Clark, 1968, p. 74)

n
The number of sample units which must be examined from a lot of food to satisfy the requirements of a particular sampling plan

Package
A unit of food within a specific container – sack, bag, box, can, bottle, drum, barrel, etc.

Population (epidemiological)
The number of people estimated to have consumed a particular food and hence representing the population at risk if the food is hazardous

Population (microbiological)
The number of microorganisms, total or of a species or group(s), dispersed within a defined quantity of food. Hence, a total microbial population or a population of staphylococci, coliforms, etc.

Population (statistical sampling)
The total number of sample units about which an inference is to be made from the found analytical results. In this book, population usually relates to the total number of hypothetical individual portions within an assignment, each identical in quantity with the defined sample unit (see page 4).

Probability of acceptance
The likelihood that a given lot would be accepted on the basis of actual test when the actual population being tested is at some stated quality level

Producer risk
The probability that a lot will be unacceptable on the basis of found values, when in reality, the lot as a whole falls within the criteria of acceptance

Psychrotrophic organisms
Those organisms, important in reducing shelf-life of many refrigerated foods, which are enumerated by incubation of test media for defined periods, in the temperature range of 0°

to 5° C, hence excluding some mesophilic bacteria whose lower limits for growth approach 5° C

Random
An adjective describing a manner of choice which excludes bias (see page 10). When applied to a sampling procedure, it implies that the procedure will cause each sample unit to have an equal chance of being selected.

Random sample
A collection of sample units obtained from the material of interest so that each sample unit had an equal chance of being selected, hence excluding bias. Reference to a table of random numbers is usually involved (see page 11 *et seq*.).

Sample
The total number (one or more) of individual sample units (ideally drawn at random) which will be tested in accord with a specific sampling plan.

Sampling plan
A statement of the criteria of acceptance to be applied to a lot, based on examination of a required number of sample units by defined analytical methods. The sampling plans defined in this book require specification as follows: (i) if a 2-class plan – n, c, m; (ii) if a 3-class plan – n, c, m, M.

Sample unit
The individual portion or container of food randomly taken as part of the overall sample and to which the analytical test will be applied.

Sanitation
Activities within a processing plant to control sources of microbial contamination, from whatever source, i.e., from air, equipment, raw foods, human contact, and time-temperature abuse that permits microbial growth. Synonymous with 'food hygiene' as used in Europe.

Severe hazard
The hazard associated with the presence of a pathogen or toxin in a food which when consumed is likely to cause severe disease, either in a normal population or in a particularly susceptible group to which the food in question is often destined. Severity of hazard is largely determined by clinical severity of the disease induced, but many ancillary factors are pertinent (see Chapter 4 and Tables 6 and 7). Most pathogens of concern in this category, such as *C. botulinum, S. typhi, Brucella* spp., *Sh. dysenteriae I* (Table 6), are normally sought for by application of investigational sampling (page 63).

Shelf-life
The period after processing during which a food remains in acceptable condition

SPC
Standard plate count – the numbers of aerobic mesophilic microorganisms present per gram in the test food sample as determined by the standard method (Thatcher and Clark, 1968, p. 66).

Stratification
A device for controlling known sources of variation. It may be used where prior knowledge exists that the consignment is potentially not of uniform quality (see page 11).

Stringency
The ability of a sampling plan to distinguish between acceptable and unacceptable lots. Stringency is increased by increasing n, reducing c, and/or lessening the microbiological criteria of acceptance.

References

AMERICAN PUBLIC HEALTH ASSOCIATION. 1967. Standard methods for the examination of dairy products (12th ed.; New York: American Public Health Association, Inc.)
– 1970. Procedures for the examination of sea water and shellfish. (4th ed.; New York: American Public Health Association, Inc.)
– 1971. Standard methods for the examination of dairy products (13th ed.; New York: American Public Health Association, Inc.)
ANDERSON, E.S. and HOBBS, B.C. 1973. Studies of the strain of *Salmonella typhi* responsible for the Aberdeen Typhoid outbreak. Israel J. Med. Sci. *9*, no. 2, 162
ASSOCIATION OF FOOD AND DRUG OFFICIALS OF THE UNITED STATES. 1969. Recommended bacterial limits for frozen precooked beef and chicken pot pies. Quart. Bull. Assoc. Food and Drug Officials US, suppl. issue
BAROSS, J. and LISTON, J. 1970. Occurrence of *Vibrio parahaemolyticus* and related hemolytic vibrios in marine environments of Washington State. J. Appl. Microbiol. *20*, 179
BASHFORD, T.E., GILLESPY, T.G., and TOMLINSON, A.J.H. 1960. Report of the Fruit and Vegetable Canning and Quick Freezing Research Association, Chipping Camden, England
BORNEFF, J. 1963. Bacteriological specifications for ice cream. Rendic. Istit. Sup. di Sanità (Roma) *26*, 475
BRAGA, A. and PALLADINO, D. 1963. Ten years of research and observation on frozen products by a large confectionery industry. Rendic. Istit. Sup. di Sanità (Roma) *26*, 468
BRAY, D.F., LYON, D.A., and BURR, I. 1973. Three-class attributes plans in acceptance sampling. Technometrics *15*, 575
BRYAN, F.B. 1971. Disease transmitted by foods (a classification and summary). U.S. Department of Health Education and Welfare, Health Services and Mental Health Administration Center for Disease Control Training Program
BURR, I.W. 1953. Engineering statistics and quality control (New York: McGraw-Hill)
BURTON, H., PIEN, J., and THIEULIN, G. 1965. Milk Sterilisation. FAO Agricultural Studies no. 65 (FAO, Rome, Italy)
CANADIAN DEPARTMENT OF AGRICULTURE, OTTAWA. 1971. Bacteriological quality of pasteurized egg products. Unpublished data
CASMAN, E.P. and BENNETT, R.W. 1965. Detection of staphylococcal Enterotoxin in food. J. Appl. Microbiol. *13*, 181
CLARK, D.S. 1965a. Method of estimating the bacterial population of surfaces. Can. J. Microbiol. *11*, 407
– 1965b. Improvement of spray-gun method of estimating bacterial populations on surfaces. Can. J. Microbiol. *11*, 1021

CLARKE, N.A. and CHANG, S.L. 1959. Enteric viruses in water. J. Am. Water Works Assoc. *51*, 1299

CLIVER, D.O. 1969. Viral infections in food-borne infections and intoxications. Edited by H. Riemann (New York: Academic Press), p. 73

COMINAZZINI, C. 1964. Microbiological specifications for the hygienic protection of milk and its derivatives. Annali Sclavo. *6*, 232

COTRUFO, P., MOLESE, A., TEDESCHI, G., and VINGIANI, A. 1957. Considerations and comments on 1103 cases of typhoid fever. Acta Medica Italica di Malattie Infettive e Parassitarie *12*, 341

CRABB, W.E. and WALKER, M. 1970. The control of *Salmonella* in broiler chickens. *In* Symposium on Hygiene and Food Production. Edited by A. Fox (London: Borough Polytechnic), p. 119

CRAIG, J.M. and PILCHER, K.S. 1966. *Clostridium botulinum* type F: Isolation from salmon from the Columbia River. Science, *153*, 311

CRAIG, J.M., HAYES, S., and PILCHER, K.S. 1968. Incidence of *Clostridium botulinum* type E in salmon and other marine fish in the Pacific Northwest. Appl. Microbiol. *16*, 553

DAVIDSON, C.M. and WEBB, G. 1972. The behavior of salmonellae in vacuum-packaged cooked cured meat products. Proc. 18th Meeting of Meat Research Workers, Guelph, Canada

DOLMAN, C.E. 1957. The epidemiology of meat-borne disease. *In* Meat Hygiene (World Health Organization, Monograph Series, no. 33, p. 11)

– 1974. Human botulism in Canada (1919–1973) Can. Med. Assoc. J. *110*, 191

DUNCAN, A.J. 1965. Quality control and industrial statistics (3rd ed.; Chicago: Richard D. Irwin)

EDEL, W. and KAMPELMACHER, E.H. 1972. Comparative studies on isolation methods of 'sublethally injured' salmonellae in nine European laboratories. Bull. World Health Org. In press

ELLIOTT, R.P. and MICHENER, H.D. 1961. Microbiological standards and handling codes for chilled and frozen foods. Appl. Microbiol. *9*, 452

FERENCIK, M. 1970. Formation of histamine during bacterial decarboxylation of histidine in the flesh of some marine fishes. J. Hyg. Epidemiol. Microbiol. Immunol. *14*, 52

FOOD AND AGRICULTURE ORGANIZATION. 1969. Yearbook of Fishery Statistics, vol. 29, pt. 1, 'Fishery Commodities.' (Rome: Food and Agriculture Organization)

– 1970. Yearbook of Fishery Statistics, vol. 31, pt. 1, 'Fishery Commodities.' (Rome: Food and Agriculture Organization)

– 1971. Trade Yearbook, vol. 25. (Rome: Food and Agriculture Organization)

GABIS, D.A. and SILLIKER, J.H. 1974. ICMSF Methods Studies. II Comparison of analytical schemes for detection of *Salmonella* in high-moisture foods. Can. J. Microbiol. *20*, 663

GANGAROSA, E.J., DONADIO, J.A., ARMSTRONG, R.W., MEYER, K.F., BRACHMAN, P.S., and DOWELL, V.R. 1971. Botulism in the United States, 1899–1969. Am. J. Epidemiol. *93*, 93

GAUB, W.H. 1946. Environmental sanitation – Colorado major health problem. A review of the problem. Rocky Mtn. Med. J. *43*, 99

GAYLER, G.E., MACCREADY, R.A., REARDON, J.P., and MCKERNAN, B.F. 1955. An outbreak of salmonellosis traced to watermelon. Public Health Rept. *70*, 311

GELDREICH, E.E. and BORDNER, R.H. 1971. Fecal contamination of fruits and vegetables during cultivation and processing for market. A review. J. Milk Food Technol. *34*, 184

GILBERT, R.J. 1970. Comparison of materials used for cleaning equipment in retail food premises, and of two methods for the enumeration of bacteria on cleaned equipment and work surfaces. J. Hyg., Camb. *68*, 221

GILBERT, R.J. and WATSON, H.M. 1971. Some laboratory experiments on various meat preparation surfaces with regard to surface contamination and cleaning. J. Food Technol. *6*, 163

GILLESPIE, E.H. 1963. Bacteriological examination of ice cream. Rendic Instit. Sup. di Sanità (Roma) *26*, 482

GORESLINE, H.E., INGRAM, M., MACÚCH, P., MOCQUOT, G., MOSSEL, D.A.A., NIVEN, JR., C.F., and THATCHER, F.S. 1964. Tentative classification of food irradiation processes with microbiological objectives. Nature *204*, 237

GROSSO, E. 1955. Five years of bacteriological investigations on the control of commercial ices. L'Ospedale Maggiore, Milano. *43*, 615

GUELIN, A. 1962. Polluted waters and the contamination of fish. *In* Fish as Food, vol. II. Edited by Georg Borgstrom (Academic Press), p. 481

GULASEKHARAM, J., VELAUDOPILLAI, T., and NILES, G.R. 1956. The isolation of *Salmonella* organisms from fresh fish sold in a Colombo fish market. J. Hyg. *54*, 581

HALSTEAD, B.M. 1965. Poisonous and venomous marine animals of the world. vol. 1 (Wash. D.C.: U.S. Government Printing Office)

– 1967. Poisonous and venomous marine animals of the world. vol. 2. (Wash. D.C.: U.S. Government Printing Office)

– 1970. Poisonous and venomous marine animals of the world. vol. 3. (Wash. D.C.: U.S. Government Printing Office)

HAMAKER, H. 1960. Le contrôle qualitatif sur échantillon. Revue de statistique appliquée *8*, 5

HARDMAN, E.W., JONES, R.L.H., and DAVIES, A.H. 1970. Fascioliasis – a large outbreak. Brit. Med. J. *3*, 502

HARMSEN, H. 1953. An epidemic of typhoid fever caused by sewage-manured vegetables. Thoughts on the agricultural utilization of waste-water of the town of Lüneburg at Bardowick. Städtehygiene *4*, 48

HECHELMANN, H., TAMURA, K., INAL, T., and LEISTNER, L. 1971. Occurrence of *Vibrio parahaemolyticus* in various lake regions in Europe in the year 1970. Die Fleischwirtschaft. *6*, 965

HERSOM, A.C. and HULLAND, E.D. 1969. Canned foods. An introduction to their microbiology. (6th ed.; London: J.&A. Churchill Ltd.)

HILLIG, F., DUNNIGAN, A.P., HORWITZ, W., and WARDEN, L.L. 1960. Authentic packs of edible and inedible frozen eggs and their organoleptic, bacteriological, and chemical examination. J. Assoc. Official Agricultural Chemists *43*, 108

HOBBS, B.C. 1965. Contamination of meat supplies. Great Britain Minist. Health Lab. Serv. Mon. Bull. *24*, 123, 145

– 1971. Food poisoning from poultry. *In* Poultry disease and world economy. Edited by R.F. Gordon and B.M. Freeman (Edinburgh: British Poultry Science), p. 65

HOBBS, B.C. and GILBERT, R.J. 1970. Microbiological standards for food: public health aspects. Chemistry and Industry no. 7, p. 215

HOBBS, B.C. and WILSON, J.G. 1959. Contamination of wholesale meat supplies with salmonellae and heat-resistant *Clostridium welchii*. Great Britain Minist. Health Lab. Serv. Mon. Bull. *18*, 198

HORWITZ, W. (ed.). 1970. Official methods of the Association of Official Analytical Chemists (11th ed.; Wash. D.C.: Assoc. Official Analytical Chemists, P.O. Box 540, Benjamin Franklin Station)

HOUSER, L.S. 1965. National shellfish sanitation program manual of operations: Part 1, sanitation of shellfish growing areas. (Wash. D.C.: U.S. Public Health Service)

IENISTEA, C. 1971. Bacterial production and destruction of histamine in foods, and food poisoning caused by histamine. Die Nahrung *15*, 109

INTERNATIONAL DAIRY FEDERATION. 1958a. International Standard FIL-IDF *2*, 1958. Methods of sampling milk and milk products (General Secretariat, 10 rue Ortelius, Brussels 4, Belgium)

– 1958b. International Standard FIL-IDF *3*, 1958. Colony count of liquid milk and dried milk (General Secretariat, 10 rue Ortelius, Brussels 4, Belgium)

– 1972. IDF Monograph on UHT milk. Ann. Bull. Int. Dairy Fed., Part V (General Secretariat, 10 rue Ortelius, Brussels 4, Belgium)

JOHNSON, H.C., BAROSS, J.A., and LISTON, J. 1971. *Vibrio parahaemolyticus* and its importance in seafood hygiene. J. Am. Vet. Med. Assoc. *159*, 1470

JOHNSTON, R.W., FELDMAN, J., and SULLIVAN, R. 1963. Botulism from canned tuna fish. Public Health Repts. *78*, 561

JOINT FAO/WHO EXPERT COMMITTEE ON MILK HYGIENE. 1957. First report (World Health Organization, Technical Report Series, no. 124; FAO Agricultural Studies, no. 40)

– 1960. Second report (World Health Organization, Technical Report Series, no. 197; FAO Agricultural Studies, no. 52)

– 1970. Third report (World Health Organization, Technical Report Series, no. 453; FAO Agricultural Studies, no. 83)

JOINT FAO/WHO EXPERT COMMITTEE ON ZOONOSES. 1959. Second report (World Health Organization, Technical Report Series, no. 169)

JOINT FAO/WHO FOOD STANDARDS PROGRAMME. CODEX ALIMENTARIUS COMMISSION. 1969. Recommended international code of practice. General principles of food hygiene. FAO, Rome

KAMPELMACHER, E.H., INGRAM, M., and MOSSEL, D.A.A. (eds.). 1969. The microbiology of dried foods: Proc. sixth international symposium on food microbiology, Bilthoven, the Netherlands, June 1968. Grafische Industrie, Haarlem, the Netherlands

KAMPELMACHER, E.H., VAN NOORLE JANSEN, L.M., MOSSEL, D.A.A., and GREEN, F.J. 1972. A survey of the occurrence of *Vibrio parahaemolyticus* and *V. alginolyticus* on mussels and oysters in estuarine waters in the Netherlands. J. Appl. Bact. *35*, 431

KAUFMANN, O.W. and BRILLAUD, A.R. 1961. Development of *Clostridium botulinum* spores in sterile milk. J. Dairy Sci. *44*, 1161

KJELLANDER, J. 1956. Hygienic and microbial viewpoints on oysters as vectors of infection. Lakartednivgen *53* (Pt. 1), 1009

KREUZ, A. 1955. Hygienic evaluation of the agricultural utilization of sewage. Genundheits-Ingenieur *76*, 206

LEININGER, H.V., SHELTON, L.R., and LEWIS, K.H. 1971. Microbiology of frozen cream-type pies, frozen cooked-peeled shrimp and dry food-grade gelatin. Food Technol. Champaign *25*, 224

LISTON, J. 1957. The occurrence and distribution of bacterial types on flatfish. J. Gen. Microbiol. *16*, 205

MELICK, C.O. 1917. The possibility of typhoid infection through vegetables. J. Infect. Dis. *21*, 28

MOSSEL, D.A.A. and BAX, A.W. 1967. Selective counting of osmophilic yeasts in food of low A_w values. Mitt. Geb. Libensm. Untersuch. Hyg. *58*, 154

MOSSEL, D.A.A., KLEYNEN-SEMMELING, A.M.C., VINCENTIE, H.M., BEERENS, H., and CATSARAS, M. 1970. Oxytetracycline-glucose-yeast extract agar for selective enumeration of molds and yeasts in foods and clinical material. J. Appl. Bact. *33*, 454

MOSSEL, D.A.A. and RATTO, M. 1970. Rapid detection of sublethally impaired cells of Enterobacteriaceae in dried foods. Appl. Microbiol. *20*, 273

NATIONAL ACADEMY OF SCIENCES, NATIONAL RESEARCH COUNCIL. 1969. An evaluation of the *Salmonella* problem. A joint report of the U.S. Department of Agriculture and the Food and Drug Administration of the U.S. Department of Health Education and Welfare, prepared by the *Salmonella* Committee of the National Research Council. (Wash. D.C.: National Academy of Sciences), 207 pp.

NATIONAL CANNERS ASSOCIATION. 1966. Processes for low-acid canned foods in metal containers. Bull. 26-L. (Wash. D.C.: National Canners Association)

– 1968. Laboratory manual for food canners and processors, 3rd ed. vol. 1. Westport, Conn.: Avi Publishing Co.

– 1971. Processes for low-acid canned foods in glass containers. Bull. 30-L. (Wash. D.C.: National Canners Association)

PIVNICK, H. and BARNETT, H. 1965. Effect of salt and temperature on toxinogenesis by *Clostridium botulinum* in perishable cooked meats packed in air-impermeable pouches. Food Technol. Champaign *19*, 1164

PIVNICK, H., ERDMAN, I.E., MANZATIUK, S., and POMMIER, E. 1968. Growth of food poisoning bacteria on barbecued chicken. J. Milk and Food Technol. *31*, 198

REAGAN, J.G., YORK, L.R., and DAWSON, L.E. 1971. Improved methods for determination of certain organic acids in pasteurized and unpasteurized liquid and frozen whole egg. J. Food Sci. *36*, 351

REPORT OF THE 1ST INT. SYMP. OF FOOD MICROBIOLOGY. *In* Annales de l'Institut Pasteur de Lille, vol. 7, 1955

ROBERTS, D. 1972. Observations on procedures for thawing and spit-roasting frozen dressed chickens, and post-cooking care and storage: with particular reference to food-poisoning bacteria. J. Hyg., Camb. *70*, 565

ROTH, L.A. and KEENAN, D. 1971. Acid injury of *Escherichia coli*. Can. J. Microbiol. *17*, 1005

SAKAGUCHI, G. 1969. Botulism type E in food-borne infections and intoxications. Edited by H. Riemann (New York: Academic Press), p. 329

SAKAZAKI, R. 1969. Halophilic *Vibrio* infections. *In* Food-borne infections and intoxications. Edited by H. Riemann (New York: Academic Press), p. 115

SCHAAFSMA, A.H. and WILLEMZE, F.G. 1957. Gestion moderne de la qualité (Paris: Dunod)

SHARF, J.M. (ed.). 1966. Recommended methods for the microbiological examination of foods. Second edition. (New York: Am. Public Health Assoc.)

SHEWAN, J.M. 1961. The microbiology of sea water fish. *In* Fish as food, vol. I. Edited by G. Borgstrom (New York: Academic Press), p. 487

– 1970. Bacteriological standards for fish and fishery products. Chemistry and Industry, no. 6, p. 193

SHOOTER, R.A., COOKE, E.M., ROUSSEAU, S.A., and BREADEN, A.L. 1970. Animal sources of common serotypes of *Escherichia coli* in the food of hospital patients. Lancet *2*, 226

SILLIKER, J.H. and GABIS, D.A. 1973. ICMSF methods studies. I Comparison of analytical schemes for detection of *Salmonella* in dried foods. Can. J. Microbiol. *19*, 475

SPLITTSTOESSER, D.F. and SEGEN, B. 1970. Examination of frozen vegetables for salmonellae. J. Milk and Food Technol. *33*, 111

STEINHAUER, J.E., DAWSON, L.E., MALLMANN, W.L., and WILKINSON, R.J. 1967. Microbial counts and certain organic acids in liquid and frozen whole eggs. Food Technol. Champaign *21*, 647

SURKIEWICZ, B.F. 1966. Bacteriological survey of the frozen prepared foods industry. I. Frozen cream-type pies. Appl. Microbiol. *14*, 21

SURKIEWICZ, B.F., JOHNSTON, R.W., MORAN, A.B., and KRUMM, G.W. 1969. A bacteriological survey of chicken eviscerating plants. Food Technol. Champaign *23*, 1066

TAKACS, J., SIMONFFY, Z., and IMREH, E. 1968. The qualification of frozen foods on the basis of microbiological investigations. Magyor Allatorvosok Lapja. J. Hungarian Veterinarians, October, p. 517

– 1969. Qualification of frozen foods with microbiological examination. Acta Veterinaria Academiae Scientarum Hungaricae *19*, 29

THATCHER, F.S. *In* Microbiological quality of foods, 1963. Edited by L.W. Slanetz, C.O. Chichester, A.R. Gaufin, and Z.J. Ordal (New York: Academic Press), p. 118

– 1969. Hygienic and safety aspects of quality control of fish and fishery products. *In* Technical Conference on Fish Inspection and Control. Food and Agriculture Organization FE: FIC/69/R.17, Rome, Italy

THATCHER, F.S. and CLARK, D.S. (eds.). 1968. Microorganisms in foods. Their significance and methods of enumeration. (Toronto, Canada: University of Toronto Press and Oxford, England: Oxford University Press)

THOMSON, W.K. and THACKER, C.L. 1972. Incidence of *Vibrio parahaemolyticus* in shellfish from eight Canadian Atlantic sampling areas. J. Fish. Res. Bd. Canada *29*, 1633

TODD, E.C.D. and PIVNICK, H. 1973. Public health problems associated with barbecued foods. A review. J. Milk Food Technol. *36*, 1

U.S. DEPARTMENT OF HEALTH, EDUCATION, AND WELFARE. 1965. Grade 'A' pasteurized milk ordinance, 1965 recommendations of Public Health Service (Wash. D.C.: U.S. Dept. Health Education and Welfare, Public Health Service)

U.S. DEPARTMENT OF HEALTH, EDUCATION, AND WELFARE, PUBLIC HEALTH SERVICE. 1968. *Salmonella* surveillance, Rept. no. 72 (Atlanta, Ga., U.S.A.: National Communicable Disease Center)

UNITED STATES FOOD AND DRUG ADMINISTRATION. 1969. Human foods; current good manufacturing practice. Federal Register *34*, 6977 (21 CFR, Pt. 128)

– 1971. Manufacture and processing of canned foods. Proposed statement of policy. Federal Register *36*, 219, 21688

VELIMROVIC, B. 1972. The geographical distribution of the human disease due to *Vibrio parahaemolyticus* in South East Asia and the Pacific. Zentralblatt für Bakteriologie, Parasitenkunde, Infektionskrankheiten und Hygiene *227*, 385

WILLS, J.H., JR. 1967. Seafood toxins. *In* Toxicants occurring naturally in foods. Publ. 1354, National Academy of Sciences, National Research Council, Wash. D.C.

WOOD, P.C. 1970. The principles and methods employed for the sanitary control of molluscan shellfish. Proceedings of the Food and Agriculture Organization Tech. Conf. on Marine Pollution and its effects on living resources and fishing, Rome, Italy

WOODWARD, W.E., GANGAROSA, E.J., BRACHMAN, P.S., and CURLIM, G.T. 1970. Foodborne disease surveillance in the United States, 1966 and 1967. Am. J. Public Health *60*, 130

Index